Peter Pan
Wears Steel Toes

Other novels by Laura Hesse

The Holiday Series (YA and Teen):

One Frosty Christmas
The Great Pumpkin Ride
A Filly Called Easter
Two Independents

General Fiction:

The Thin Line of Reason
Gumboots, Gumshoes & Murder

Non-Fiction:

Peter Pan Wears Steel Toes

Peter Pan
Wears Steel Toes

By Laura Hesse

Running L Productions

Parksville, BC

Peter Pan Wears Steel Toes

National Library of Canada Cataloguing in Publication Data

Hesse, Laura - 1959

Peter Pan Wears Steel Toes/by Laura Hesse

ISBN 13 digit: 978-0-9877343-3-4

Publisher: Running L Productions
Parksville, BC Canada
http://www.runningLproductions.wordpress.com

ACKNOWLEDGEMENTS

I want to thank all my cruising partners and supervisors over the years, wherever in the world or the universe you may be. This includes but is not limited to Brian, Jim, Denise, Dave, Nancy, Hilda, Micheline, Ben, Joanne, and Paul. It was a pleasure working with you. My life is much richer for the experience.

A big thank you goes out to Boris Rasin in New York who did an amazing job on the front cover. He took my vision and made it a reality.

Thanks to my editor, Dianne Andrews, for a job well done.

I'd also like to thank my mum for squirreling away this manuscript. I had thought it lost forever.

A big thank-you goes out to my family for supporting me throughout my life and various incarnations.

FROM THE AUTHOR

I wrote the original manuscript for this novel during the winter of 1982 on an old Underwood typewriter while the memories were still fresh in my mind and the daily journal that I had kept was still around. It was happenstance that I found the manuscript stored inside a box of my mother's papers several months ago.

Reading it again brought back memories of an amazing journey: gruelling days of hard work, wonderful people, life experiences and adventures galore.

Everything that happens in this book is true. The adventures all occurred during the summer that I spent working for the Ministry of Natural Resources in Blind River. Names and characters have been changed in the interest of privacy. Any resemblance to real persons, living or dead, is purely coincidental.

Please remember that this book is set in 1981. You may not agree with some of the language contained herein. Canada was also making the switch to metric and we were recording distances and measuring plots in both imperial and metric which is why I jump between the two.

CONTENTS

Chapter One

Careful What You Wish For

Blind River, April, 1981

'No Whites Allowed' was scrawled in thick white acrylic paint across the rusted supports of the train trestle that traversed the river on the border of the local Reserve and Blind River's town boundaries. These sentiments were out of my realm of experience.

One has to understand, I was privately adopted from the Salvation Army Hospital in Ottawa by two wonderful people. My father was a proud French Canadian whose mother was of Mohawk heritage and whose father was Belgian. My dad worked his way up from a government clerk to a Certified General Accountant, becoming the Chief Financial Officer of the Royal Canadian Mint by the time I was seven. My mother was an English war bride who taught dancing and choreographed and produced music hall theatre.

Both survived a war that shredded countries and lives. They buried friends and loved ones. They never discussed World War II, except to share enchanting stories of laughter, music and song, as seen through the eyes of love and youth.

My brother, almost eleven years older than me, was and is a city boy at heart. I'm sure he still wonders how a horse-addicted tree-hugging tomboy grew out of the cute little baby that first arrived on the Groleau doorstep in the summer of 1959.

The only thing that I knew about my birth mother was that she was short, dark skinned and raven haired, most likely of Native or Métis descent, and that my birth father was tall and blond. My adoption was arranged by my parent's pediatrician because he worked at the single mother's clinic at the Sally Anne's hospital and knew that my parents had always wanted another child. He thought that my mother and I were a perfect match as she was short and dark haired and I was expected to be the same. I was adopted two months before I was born. I consider myself one of the luckiest daughters and sisters in this world.

In a profound way, those three short words written on that train trestle so many years ago signaled the beginning of a new life for me, a life so different from the one that I grew up with that it couldn't help but shape me into a whole new person. I found racism and intolerance in places, awe and respect in others. Laughter and fear, friendship and adventure, became a part of my daily life.

While the journey originally began at the Ministry of Natural Resources in Lanark, Ontario in 1979, the real challenge began in Blind River in late April of 1981. From there, I went north to Fort McMurray to work as one of the first woman firefighters hired by the Alberta Forest Service, but that is another story.

Blind River in nineteen-eighty-one was the same as most small highway towns, a little seedy and rundown but, underneath it all, populated by a friendly and charismatic group of tree-huggers, laid-off loggers, mill workers, old miners, outfitters, and mom and pop small business operators.

The grocers sold over-priced meat, milk and eggs.

The gas stations sold gallon buckets of fat, juicy worms along with fishing and hunting licenses.

The outfitters took Davey Crockett want-to-bes into the woods to shoot anything that moved during hunting season and also picked up the pieces of the headless, skinned bears left to rot in the dump by the armchair warriors too lazy or broke to hire them.

I didn't own a rifle…yet, and preferred to fish from a canoe.

In the morning, I was scheduled to start my new job with the Ministry of Natural Resources, formerly called the Department of Forests, aka, the Department!

I was never sure if that made me a Departmenter or just Departmental.

I waved goodbye to the good-natured Greyhound Bus driver who had helped me move a gargantuan, sleeping drunk off my numb shoulder in order to step off the bus. The Greyhound pulled back onto the highway in a puff of greasy diesel smoke, heading east to Sudbury.

I sniffed the air, my nose wrinkling. The smell was noxious.

It wondered if I shouldn't have taken a summer waitress job at home in Ottawa instead of trying to save the world one tree at a time.

After a moment, I realized that it wasn't the Greyhound bus or the Town of Blind River that I was smelling, it was me. I smelled of stale cigarettes, Lucky beer and Five Star whiskey.

I shouldered my backpack and hunkered down against the bitter wind blowing in off of the Great Lakes.

The sunset over the river was breathtaking. Reds, oranges and deep violets burned in brilliant plumes across the blackening sky. The rainbow of color almost erased the squalor of the dilapidated buildings beneath it.

After making my way from one full motor hotel to the other, the blinking 'No Vacancy' signs hurt my eyes. They were as red and raw as my nerves.

I stopped, breathless and tired.

My head hurt.

My feet hurt.

I wanted nothing more than to be a cat in front of a fire.

I shuffled my pack to the other shoulder and made my way into town as the last rays of the sunset reflected back at me from empty shop windows. The momentary beauty gave way to a feeling of desolate isolation.

There was no one to call, no couch to surf.

I had no friends here.

No family.

The whine of a slide guitar caught my attention.

Country music blared from an old lady of a hotel on the corner.

The Algonquin was an ancient establishment that still sported Men's and Ladies entrance signs. The hotel had seen better days, but at least the 'No' in front of the vacancy wasn't lit up. Go figure.

I had a hundred dollars in my pocket. It needed to last me until my first pay cheque and that was four weeks away. Being picky wasn't an option.

As I looked up at the faded, blistered paint and the two boarded windows on the third floor of The Algonquin, a park bench seemed like a better idea.

I steeled myself and opened the door to the Ladies' Entrance; after all, I was a lady even if I didn't look or smell the part.

Threadbare oriental carpets covered the floors of the lobby. Faded black and white pictures lined the walls. Ghostly faces watched me walk up to the counter. I felt the ladies starched disapproval and the men's scowls upon my back.

The air smelled like mold and cheap whiskey. A grand chandelier, though dust and cobweb covered, lit up the room. Inside the bar, the lead singer did a good imitation of Johnny Cash and Willie Nelson wailing about jail birds and lost loves.

A grandmotherly type of woman with graying hair and a red swollen nose puffed on a mentholated cigarette behind the front desk. She eyed me keenly.

"I'd like a room for the night, please," I asked sheepishly.

"Are you sure, honey?" she replied, kindly. "Maybe you'd like to see one first?"

"Everyone's full. I need a room."

"Okay, but there's no TV and you have to share a bathroom. Bath's at the end of the hall."

"That's fine," I squeaked. "How much?"

"Seven bucks."

"Alright," I said, taking out my wallet. I counted out seven one dollar bills. At least, it wasn't going to break the bank.

The clerk gave me a key.

"Take 301. You won't hear the bar as much and don't...I mean DON'T open the door for anyone," she warned.

"Yes, ma'am."

"By the looks of the steel toes you're wearing, I take it you're going to work for the Department?"

"Yes, ma'am, I start tomorrow."

"There's another girl here too. She's in 306, but I haven't seen hide nor hair of her so I expect she's already hunkered down for the night. In the morning, the chink joint next door has good and cheap-as-a-dollar-whore breakfasts."

"Thanks," I said, signing the register and heading up the narrow stairway to the third floor.

I found 301 and sidled inside with a backwards glance over my shoulder. A chill swept over me, my spine tingling, despite the dusty hot air blazing out of the iron heat registers in the floor.

"Leave it to me to pick a haunted hotel," I muttered to myself as I locked the door behind me.

The room wasn't quite as bad as I expected, but not great either. Paint peeled off the walls beside the door and behind the bed. The floors were clean, the sheets on the bed white and crisp. The two blankets on the bed only had a couple of small cigarette burns. The air smelled like dirty laundry, but 'beggars can't be choosers' as the old saying goes.

I decided I'd feel better sleeping on top of the bed in my own sleeping bag than under the covers and rolled out my bedroll. I set my small travel alarm for 6:30 am and changed into my flannel pjs. I've always adored the feel of flannel against my skin and the lemony smell of Sunlight reminded me of home.

I had to pee like a racehorse, but the idea of peeing in a shared bathroom didn't really do it for me. I decided I could wait until morning and dreamed of flash floods and rolling waves all night long.

5

❧ ❧ ❧

The alarm went off with a loud clamor.

I stumbled out of my sleeping bag, remembering at the last minute to change into my well worn jeans and lumberjack shirt before dashing down the hall to the community john.

The door was locked.

My kidneys were screaming. I squeezed my legs tightly together.

I waited…and waited.

I pounded on the door.

"Coming," came the muffled call from behind the door. "Hold your water."

The door opened.

A short, heavy set brunette with glacial blue eyes and an infectious smile blocked the doorway. She was dressed in jeans, flannel lumberjack shirt, and steel-toed boots, but the similarity stopped there. She was dark to my light, broad faced to my china-doll round, size sixteen to my size seven.

"Bursting, are you?"

"Yeah," I squealed, dashing past her.

She howled with laughter as I slammed the door in her face.

"You heading for the Ministry?" she called.

"You mean the Department?" I warbled back, my bladder cooing with relief.

"One and the same. Want to meet in the lobby?"

"Sure."

"The Chinese joint next door has pretty good food. We can grab some breakfast."

"Yeah, I heard that. Be right there," I called.

"My name's Sandy."

"Laura."

"See you downstairs."

"Okay."

After a quick top-to-tail wash with paper towels, I returned to my room, rolled my bedroll, and packed up my gear.

Backpack over my shoulder, I met Sandy in the lobby.

"Cruiser or compassman?" Sandy asked as we headed out the door.

"Cruiser. You?"

"Compassman. Maybe we'll get teamed up together."

"Cool."

We grabbed the last empty table in the restaurant. The waitress came and we quickly ordered the bacon and egg special, not knowing when we'd get a chance to eat next.

"It'd be great working with a girl. No tan lines."

We both laughed.

"Not likely," said the guy in the next booth over. "They usually pair guys and girls together so the man can handle the heavy lifting. Some of the canoes are heavy and the portages long. You girls need help with that."

Sandy rolled her eyes. I snorted coffee into my sleeve.

"I couldn't help over-hearing."

"Not when you're eavesdropping," Sandy quipped, openly hostile.

"Yeah, well, sorry. My name is Jimmy. I start today too," a red-faced Jimmy replied.

Jimmy was handsome in a first string quarterback kind of way. He had jock written all over him. Sandy hated him instantly.

"Laura," I reached over to shake his hand, "but my friends call me Lolly." Sandy glared at me.

"Why Lolly?" Jimmy chuckled, shaking my hand.

His hands felt dry and callused.

"Short for Lily Lolly Lou. My friend's dad used to call me that and it stuck."

"Lolly. It suits you."

"Thanks," I shrugged good-naturedly not knowing that the shorter LOL would end up the much beloved Laugh Out Loud. I like to think I was the forbearer of that expression.

"Sandy," Sandy grumbled under her breath as our breakfast arrived.

Jimmy slipped into the seat beside me.

"Mind if I join you?"

"You already have," Sandy quipped.

"Thanks," he replied, not noticing her sarcasm.

Jimmy seemed to be the stereotypical jock. He neither noticed my silence or Sandy's quick witted responses to any question he asked from the status of the weather to the chances of the Ottawa Rough Riders winning the Grey Cup. He was an Argo fan all the way.

After the third round of Jimmy's harping on women's bodily functions, I was beginning to understand what Sandy had already fathomed. Jimmy was a bit of an idiot. A summer with him might indeed drive me mad.

"You know you guys have to be careful in the woods when you're on the rag. The bears can smell it for miles," Jimmy said through a mouthful of ketchup covered scrambled eggs.

Except for the bacon, I had suddenly lost my appetite.

"Noooo," said Sandy, spearing a hash brown. "Why is that?"

"Cause, well, you know, the smell of blood," Jimmy shrugged.

"What does the smell do to them, Jimmy?"

"It gets them, you know, randy," he said shyly. "The bull mooses too. You girls got to be careful."

"Fancy that, we might get gang banged by a horny bear or a bull moose, Lolly."

"Sandy!"

Jimmy blushed like a cherry blossom, stuttered and stood up, knocking the ketchup and coffee cups across the table. His face grew an even darker shade of red. He stuttered and then bolted from the table.

"That was mean," I whispered to Sandy, sopping up my spilled coffee with a napkin.

"Yeah, I guess."

"He might be your partner," I chided. "Go apologize."

"I will not!"

"Yes."

"No."

I pushed away from the table and hustled over to the counter to pay my bill.

"Jimmy, why don't you wait for us and we'll walk down to the Department together," I said, already feeling like a local.

"Are you sure?" he whispered. "I don't think Sandy likes me."

"She'll get over it," I said encouragingly, if not half-heartedly.

Jimmy wasn't all ego and his intentions were for the best.

We paid our tabs and then hoisted our fifty plus pound packs over our shoulders. The packs contained all that we needed for the next four months. In my case it was five pairs of undies and socks, two t-shirts, one extra pair of jeans, one set of shorts, one sweater, one flannel nightie, toiletries, a half empty box of Tampax, and Ibuprofen. On the back of the pack was fastened a sleeping bag, deflated air mattress, and a raincoat.

🦢 🦢 🦢

In the nineteen-eighties, it seemed that all Ontario small town government buildings were painted white with dark green trim and Blind River's was no different than Lanark's had been. There was a main office, a two storey cluttered affair, and a couple of office trailers used as over-flow offices and a training centre. Later in the summer, I decided that these trailers were to keep us out of the main building. With a schedule of ten days in the bush and four days out, we stank like dead skunks by the time we finished the ten day stint. I've smelled a dead skunk and, baby, we were close.

The biggest building of the bunch was the shop and garage. The compound behind the shop was filled with beat up 4x4's and canoes, with the occasional small motor boat and trailer. The trucks, canoes and boats were no more than two years old, but they looked like they had endured thirty years of hard use.

A tall taciturn man of about thirty-two waved to us from the steps of the main building. It wasn't hard to tell the cruisers from the other Department workers. Our uniform was Grebb steel-toed boots, lumberjack shirts, and jeans. Bright yellow hard hats swung from our backpacks.

The yellow hard hats were a prerequisite. All worker ants wore them. The orange and red hard hats designated the crew leaders and queen bees of upper management.

Ben Wilson, our Team Leader, sometimes Guidance Counselor, part-time friend and mentor, escorted us to the conference room where numerous other cruisers were milling about, introducing themselves to each other.

"Okay, grab a seat everyone," Ben called over the hubbub. "We've got lots to cover."

Chairs scraped across scarred linoleum as everyone tossed their packs into a corner and sat down. Some nervously fidgeted with their pencils and paper in front of them, while others leaned back nonchalantly, clearly bored by having gone through this several times before.

"First order of business is to get you to fill out the contact emergency information forms. Don't forget to write down any allergies you might have or meds you take so we have that on file in case we need it."

"Like if you set yourself on fire with a Coleman stove," one cruiser offered.

Everyone chuckled, thinking this enormously funny. Who on earth would think of a career in forestry without a love and knowledge of the outdoors and camping?

"Don't laugh," Ben said. "Sean set his tent on fire last year."

"Seriously," Sandy blurted out.

"Seriously," Sean said. "I got second degree burns slapping at the tent to put it out. I know, it was dumb, but it was an instant reaction. I didn't think."

Sean held up his two scarred hands. Everyone cringed.

"That's why we do a one week orientation course on first aid, compassing, tree species identification, and the SAFE operation of camping equipment," said Ben.

There were a few snickers around the table.

"Don't snicker and don't laugh. You think college prepared you for real life, you are sooooo wrong. Last year, we had two guys fail to report in. We found them lost, two hundred feet in the bush from their camp. They thought their compasses lied to them."

Silence greeted his words.

"But they did the right thing. They stayed put and waited for help. There is lots of magnetite in this area and it will screw with the compasses so be careful and pay attention in orientation. This isn't a camping party with your mother or father. That is why we send you out with radios. You will report in EVERY night to base camp. If you don't report in by morning, we come looking."

A chill swept over the room.

I looked around. Several of the girls and guys looked glum. Sandy made a clown's face at me. Jimmy puffed up like he had just consumed a Christmas turkey all by himself. The seasoned cruiser, whose partner had set himself and their tent on fire, looked smug. A compact, bullish fellow with a crew-cut seemed delighted. His name was Colin and he set my alarm bells to ringing.

There was one girl, Ocean, who stood apart from all the rest. She smiled sweetly at everything that Ben said and wrote copious notes. She had an air of purity about her, from the delicate poise of her pen to the soft timbre of her voice. She was pretty in a country maiden sort of way. I learned later that she had been raised on a commune on some island off the west coast.

I realized that I didn't want to be paired with any of them except for Ocean, and secretly hoped that Ben was up for grabs.

"The other thing we are going to talk about right now is bears."

There was a series of groans from the women. Yeah, be careful when you are on your period, we thought. Yadayada.

"Do not yell, wave your arms over your head, or bang pots at them. That is just stupid. If you run into a black bear, freeze and wait for a moment. If he doesn't know you're there, because they do have bad eyesight and hopefully you are downwind, then back

11

away slowly. Once you are out of range, put on your Peter Pan shoes and fly out of there."

This was met with a chorus of laughter.

"If you are attacked, roll yourself into a tight ball covering your head with your arms. It isn't about playing dead, it's about covering your vital organs and your head is one of them. Usually, the bear will bat you around a bit, maybe try and cover you up with leaves, and then wander off. It's been years since anyone has been killed by a bear around here so let's keep it that way. Keep your food covered and away from your tent. Alright, it's time to pick crews and issue equipment," Ben finished.

The two man crews were composed of one compassman and one cruiser. There were six crews in all. Each crew was assigned a compass, computer data sheets, camping equipment, a haga or sunto (an instrument used for measuring tree heights), an increment borer for taking tree age readings from the rings, and a prism (an instrument for reading trees of over a given basal area at a specific geographic point).

In layman's terms, we counted trees. We measured stand density and health. These records helped with forest regeneration surveys, woodlot allotment, and planning.

It was a lucky day for me. Ha-ha. Sandy and I were paired together. We were the only all female team.

I thought Jimmy was going to throw himself down on the floor and give us twenty-five when Ben partnered him up with Debbie.

Debbie made frayed denim jeans and scarred leather work boots look like Gucci and Prada. She was a bean-pole blond with laugh lines and a winning smile. She had the unique talent of making everyone around her feel at ease.

I liked her.

Big surprise, Sandy did not.

"I need everyone to hand me their paperwork. We're going to head out to the first trailer on the left to pick up your area maps. We'll assign you your trucks as well," Ben yelled over the din. Each team signed off on the their equipment.

"This way, guys," he said, collecting the last of the paper work and escorting us outside.

Sandy hung back.

"Come on, Sand, let's go see where they're sending us?"

"We've still got a week. What's the rush?"

"What's your problem?" I asked, annoyed.

"Nothing," she replied, eyeing Debbie critically.

"What? Debbie? What do you have against her?"

"Oh, come on," she said, dragging her boots across the floor as we followed everyone else outside. "It's just, did you see her nails? They are frigging perfect. Everything about her is so frigging perfect."

A light bulb went off.

"Did you hope you were going to be paired with Jimmy? I thought you didn't like him the way you were acting."

"Nooo," she whined.

In the compound, Ben hurried the other crews into the trailer. He stood patiently waiting for us.

I shrugged helplessly at him.

"Girls?" Ben asked as Sandy ducked under his arm and carried on into the trailer.

"Everything okay, Laura," he whispered in my ear.

"No sweat, Ben," I replied smoothly, hoping he didn't hear the tremor in my voice. Was it okay? I was at a loss.

Inside the trailer, Ben handed out the keys to the trucks and maps of the areas that we were each assigned to.

Because I was the cruiser and senior member of a two man crew, the truck was issued in my name along with a temporary government license. Sandy was given one too, but ultimately the return of the truck and any equipment was mine.

"Laura!"

"Yes, Ben," I chirped innocently, waiting for the hammer to fall.

"We're short on rentals so I'm giving you one of the Ministry trucks. Don't make me regret it."

"S'no problem," I replied.

13

"Cruiser arrested for drunk driving," Colin snorted in disgust. "We'll take it, eh, Ocean?"

Ocean looked uncomfortable. She had already been paired up with Colin.

"Not happening,' I replied angrily, instantly feeling sorry for Ocean as the tension in the room was palpable.

I didn't add that the year before when I was working in Lanark, my partner and I parked our truck at the far end of a local pub's parking lot in order to do our stand surveys in the provincial forest that backed onto the pub. We had a five hour line to do and it was too dangerous to park the truck on the highway. We never thought about the government sticker on the side of the truck. The big boss called us into the office when we got back because of all the phone calls.

Red-faced we showed the boss our map and aerial photos of where our cruise lines were in relation to the pub as well as all our data sheets. He leaned really close when he examined them, but our sheets clearly showed that we had in fact cruised them.

Lesson learned.

"Where are we doing our training?" Jimmy asked, changing the subject.

I smiled gratefully at him and he winked back.

"At our base camp. It is twenty minutes north east of Elliot Lake in Mississagi Park. There's a road map with your cruising maps if you get separated from the group. The camp is circled. Easy-peasy."

"Enjoy it, people, it's our last week with indoor plumbing," Debbie joked.

"Let's get packed, folks. It's an hour drive and we have lots to do," Ben urged.

"I have to stop at the store first, Ben," Debbie chirped. "I need tampons. I know you're all going to need them too. Am I right, ladies?"

Every girl smiled. Every guy groaned.

"Thanks," I whispered to Debbie on the way out, "I didn't know how to ask."

"Up front and with gusto, girlfriend," Debbie added with a grin.

"I like your gusto, Deb," I responded.

"Laura," I said, shaking her hand.

"Friends call her Lolly," Jimmy teased.

"Lolly, it is," Debbie agreed.

Sandy stalked by me, heading for the parking lot.

"What's got her knickers in a twist now?" Jimmy inquired.

Debbie roared with laughter.

"I like that. Knickers in a twist. It's very visual."

"No idea. Not going to try to figure it out. See you guys at camp," I offered, rolling my eyes in Sandy's direction before chasing after her.

"Just PMSing," Debbie said to Jimmy. "Get used to it, sunshine, you have to live with me in a tent for four months."

Fifteen minutes and a Tampax stop later, we were on our way.

🦅 🦅 🦅

Sandy and I were the last truck out of the yard. I let Sandy drive, hoping that riding around in a sage green truck with a Ministry of Natural Resources emblem emblazoned on the side would cheer her up.

"Sandy," I cautioned her as we pulled out of the drugstore parking lot, a long way behind everyone else. "Remember what Ben said."

"Yep," she said, her eyes devilishly bright. "But he said it to you."

She peeled rubber at the stop light and gunned the truck, engine revving loudly, down the highway. The speedometer climbed to seventy miles an hour.

"Not cool," I screamed at her as we barreled down the road. Luckily traffic was light.

Twenty minutes out of town, OPP sirens peeled behind us and lights flashed.

I wanted to be sick.

Sometimes trying to do the right thing really sucks.

"Sandy, pull over," I said, the ice in my voice brooking no arguments. My nerves were frayed.

"Watch and learn, sister. We're up a creek if we get a ticket."

The cop who pulled up beside us was pissed. He motioned for us to pull over.

Sandy leaned sideways.

"I can't! I can't! The accelerator's stuck," she screamed at the cop.

"What," I whispered sharply.

She flashed me a warning look and rolled down her window. She screamed again.

I had visions of being taken away in handcuffs and never working for a government agency again.

"It's stuck!"Sandy pointed down at the accelerator. "What do we do?"

Understanding blossomed on the cop's face. He made pumping motions with his hand, his face turning from anger to concern in a split second.

Sandy shot him a quick thumbs-up sign.

"Are you out of your ever loving mind?" I muttered to my crew mate.

Sandy faked pumping the accelerator and then the brakes. She slowed down and pulled over to the shoulder, the cop car pulling over in front of us.

"Just follow my lead, Lolly," she grinned. "We are friends, right? I can call you that?"

I didn't have time to answer.

With one foot, Sandy slid her jacket out from under the seat and onto the floor by the accelerator pedal.

The cop walked over to the truck. My hands shook when he opened my door.

"You okay, Miss?" His voice was steeped in concern.

"Yeah, thanks," I croaked, misery creeping into every fiber of my being.

"Oh, officer, the accelerator got stuck and I was afraid to pull over and I didn't know what to do and if I hurt this truck, I'll get fired and,…" Sandy sing-songed.

"S'okay. You did the right thing."

"My jacket fell off the seat and I reached down to pick it up, but it got stuck," she cried.

The cop shook his head.

"License and registration," he asked us.

I pulled out the registration from the glove compartment and Sandy handed over her Ministry license.

I stoically sat in my seat, my lips sealed as if I were in Judge Judy's court, every foolish comment my accuser made a point in my favor.

"You aren't going to give me a ticket are you? Seriously, I'll lose my job," Sandy said, her lower lip quivering.

"Just roll up that jacket and put it behind the seat and slow down, young lady," the officer said, handing back the registration papers and Sandy's license. "I'll let you off with a warning, but be more careful. You ladies could have been killed."

"I will, I will."

"Thanks, sir," I nodded at the officer.

He hitched up his gun belt as he walked towards the cruiser.

I sighed with relief.

"Like I said, just follow my lead," Sandy grinned and turned the key in the ignition.

I wanted to turn off the engine, but was too shaky to take over driving.

Sandy pulled slowly out onto the highway, the cop tucking his cruiser in behind us. We waved good-bye to the officer. He raised a hand and waved, then followed us down the highway for another twenty minutes.

When the cruiser was out of sight, Sandy put the pedal to the floor and we almost missed the turnoff to Elliot Lake.

As we headed up the last winding curve towards the camp, I started to giggle. I couldn't help it. Laughter bubbled out of me like skittles in soda pop.

Sandy started to laugh too.

"Oh, no," Sandy said, tears streaming down her face.

"What?"

"I have to pee."

"Me too."

Sandy pulled over. We leapt into the bushes at the side of the road. As we dropped our drawers, Ben's Ministry of Natural Resources truck pulled up beside ours.

"We are so busted," I cried.

"Are we done yet, girls?" Ben said, leaning out the window. A smile crinkled his lips.

Sandy's head popped out from behind a small spruce tree. I pulled an alder sapling over sideways to cover my bottom and peeked around the scruff of leaves.

"Duh, give us a minute, Ben," Sandy scolded.

Ben laughed and shook his head.

"I didn't think I'd have to be rescuing one of my teams this early in the game. You have something to tell me?"

"Yeah, Laura's got a kidney infection. This is our twelfth frigging stop."

"I do not," I whined.

Ben waved us away.

"Camp's fifteen minutes up the road. See you there. Have either of you seen Jason and Debbie? They aren't here yet either."

"No," we said in unison.

Ben disappeared up the road.

"See, we aren't last. Stop worrying."

"By a tortoise's breath on a hair's ass," I said.

This prompted another fit of giggles.

We drove up to the base camp. The camp was picturesque. A large cook house and several smaller bunkhouses overlooked a small lake with docking facilities. We parked in the employees parking lot beside the other trucks lined up in a row outside of the bunk houses.

The lunch bell was ringing so we made our way directly to the cook house.

Ben sauntered into the dining room with Jason and Debbie in tow half way through supper.

Jason got a speeding ticket.

Sandy and I wondered if it was the same cop that stopped us and unanimously decided not to discuss it.

I felt an ulcer coming on.

Chapter Two

Training Daze

After a couple of days of training and orienteering, we were ready to do some sample test plots which included a two mile compass traverse through the bush.

Ben started the crews ten minutes apart with the intent that we would all meet up at the end of the course.

Ah, the best laid plans of mice and men.

Ever play a game of telephone when you were a kid? You know the one. 'A barrel full of monkeys' ends up 'Barney's blowing chunkies' at the end of a chain of whispering the first line into a wide circle of adolescent ears.

Imagine setting six sets of two person teams into the bush with just a compass and a map and then expecting them to come out at the same point at the end.

Let's face it. Some have short legs. Some have long legs. Some have boundless energy when crawling through bramble thickets and falling over stumps, while others like to meander and pick dirt from under their fingernails at regular intervals (not all being women, by the way).

It was a free-for-all.

Sandy and I were Crew Six. Jimmy and Deb were Crew Five.

Jimmy was the cruiser and Deb was the compassman. They started ten minutes in front of us.

In our case, I was the cruiser and Sandy was my compassman.

"Okay, girls," Ben said to us. "See you at the finish line."

"You bet'cha," I agreed, wondering why he was singling us out when in the distance, a chorus of "Hey, Ocean, where are you?", "Shit!" and "Damn" could be heard echoing through the forest.

I watched Ben jump into his 4x4 and drive away with a knot in my stomach.

Sandy examined the map, took a compass bearing and waited. And waited. And waited.

In the distance, I heard Jimmy shout at Deb.

"Are you sure you know where you're going?"

"Of course, I do, I'm just following the footprints," she yelled back.

"That doesn't mean they know where they're all going," he called back.

"Then at least, we'll all be together," she shouted brightly.

Sandy looked up at the cloudy sky.

I knew she was stalling.

"Did you bring your rain gear," she asked.

"Yeah, it's in my pack," I responded. "Think it's going to rain? The clouds are pretty high."

"I think we're going to be out here a long time and by the time we finish, it's going to be pissing cats and dogs."

"Are we going to follow the footprints too?" I joked.

"Not on your life," she commented dryly.

"Let's rock and roll then," I offered with a grin, wondering if I really should have been that happy to be paired with Sandy. There was something 'off' about her, but I couldn't put my finger on it.

Sandy smiled back, and we set off.

It's a strange feeling walking into miles and miles of open country, with nothing but trust in a little circular black instrument with a plastic face and tiny red needle pointing north to guide you were you're going.

One missed step can leave you with a broken leg or ankle, stranded five miles or more through brush and thickets from the nearest bush road or a hundred miles from the nearest doctor or hospital. Your life depends upon your partner's ability to get you there and that little red needle.

Two years earlier, I worked Creel Census in Lanark, a small town an hour's drive from Ottawa. I basically travelled around in a boat all day, admiring the cottages along the shore, talking to fisherman and measuring the fish they caught for length, weight and health to help determine fish stocks. It was fun and I went home at night to a hot shower and a warm bed.

I wasn't in cottage country anymore.

I followed Sandy into a thick stand of poplar and birch. The ground beneath our boots was black and muddy. There were still patches of dirty snow in the dark hollows under the thorn laced thickets and around the base of spruce and balsam trees. The nights were bitter cold with temperatures dipping to the freezing mark.

Rivulets of water ran down the hills to the north of us.

The air smelled crisp and fresh, but had an under-lying odour of damp earth and mud.

"This way," she said, confidently.

Sandy didn't fool around. Her certainty relieved a lot of my anxiety. I realized suddenly that maybe I was glad that I was paired up with her. She seemed to know the bush.

Here and there, boot prints from the crews ahead of us could be seen in the swampier areas. They criss-crossed each other at odd angles and uneven intervals.

Through it all, Sandy maintained her bearing, stopping every so often to consult the all-terrain map.

A deep hole in the muck and a socked footprint showed where the mud had sucked the boot right off of someone's foot. I tried to remember if I put a spare set of socks in my day pack?

We ducked under the bows of a stand of white pines. Thick beds of needles scraped along our hard hats. Pine scent wafted over us. It was a beautiful moment. I thought of my mother trimming the tree last Christmas.

The only sounds we heard were those of our labored breathing and the chirp of birds.

We stepped out of the woods and into a thicket of brambles along the banks of a small stream. Water cascaded in a small waterfall over a broken old log.

Deb sat on her tush on the other side of the stream, draining the water out of her boot. Jimmy hovered over her, concern on his face.

Deb laughed when she saw us.

"Darn log broke underneath me," she said.

I smacked Sandy in the arm before she made a smart crack. She just rolled her eyes at me.

"Are you okay?"

"Yeah," Deb said, looking up at me.

Deb tugged on her wet boot and stumbled to her feet. Jimmy steadied her with a ready hand.

"Oh, this is gross," she said, wrinkling up her brow. "It's like walking in jello."

"I bet," Sandy quipped.

"You guys know where you're going?" Jimmy asked.

"Deb thought her compass was wrong when the other crews went farther east than where we are now."

"What did Ben keep stressing to you in our meetings?" Sandy asked, impatiently.

"Always believe your compass," Deb and I said in unison.

"Right!"

"But what if there's magnetite underneath us and it's affecting our readings," Deb asked, her voice trembling.

"You only have to worry about magnetite around the cliffs and on fault faces," Sandy counseled.

This time Sandy was more patient.

"You aren't going to have any footprints to follow next week, Debbie," Sandy said quietly. "Do it now, on your own."

"You're right," Deb sighed.

"You guys going to be okay?" I nodded to Jimmy.

"We'll be fine," he responded.

"Thanks, guys," Debbie said, squaring her shoulders. "You're right, Sandy, I do have to do this on my own."

"It's your job," Sandy reiterated.

I waved over my shoulder as Sandy and I struck out again, following the path the compass defined.

Behind us, I heard Debbie say to Jimmy, "We're going this way."

"Yeah, but that's in the opposite direction of everyone else's tracks and way off from where Sandy and Laura are going."

"I don't care. This is where the compass is pointing and that's where we are going," she demanded.

Sandy and I continued on our course.

"You think we'll ever see those two again?" I asked Sandy when we stopped for a water break.

"Maybe. Maybe not."

"Looks like we're on the right track at any rate."

A gully yawned open before us. Down in the gully, saplings were snapped in half and snow banks were kicked apart by uncaring feet.

"You ever been lost out here?"

"No," Sandy said, after a moment. She looked me in the eye. "I worked here last year, but in the office. You?"

"No. I worked Fish & Wildlife in southern Ontario. You could always hear a highway."

We put our water bottles away and climbed down into the gully.

"I thought you worked in the bush last year."

"I manned the radio and did the night service, plus checked and re-checked everyone's reports."

"By the way you talked, I thought you were old hat at this."

"Next week, we really are going to be on our own. I wasn't kidding when I told Deb that we have to take this seriously. There are no roads where we'll be."

I didn't know whether to laugh or cry.

On the plus side, Sandy's honesty was appreciated and it was nice to realize that the wisecracker attitude was just a front.

We continued on in silence after that.

Sometimes we followed other teams' tracks and sometimes we made our own.

I pondered over what it was to be lost in the bush, especially on a fly-in where we were hundreds of miles in the bush.

Would I panic?

Would I be able to stay in one place or would I keep trying to walk out?

What if we got lost on the second day of our ten day stretch and the weather turned making rescue impossible? Could we survive?

What if we lit a signal fire and the jack pine and balsam fir ignited, turning the world around us into Dante's inferno?

I pictured myself squatting by a stream, filling my hardhat full of water, and turning around to throw it on an ever-growing avalanche of flames.

I puked.

"Shit, girlfriend, you okay?"

Sandy rushed back to me.

"Drink some water. You're overheated."

She was right. I was burning up. We had been crawling up hills and down gullies for three hours already, wading through marsh and scrub brush. I was dehydrated.

Wow! It crept up fast!

I opened my pack and grabbed a fresh water bottle. I swished my mouth out and then polished off the rest.

A spattering of raindrops fell from the clouds.

Sandy and I donned our yellow rain gear.

"Sorry about that," I offered.

"Just tell me if you need to rest for a moment," she said, concerned.

"And think of how buffed I'll be at the end of the summer."

"And no tan lines."

We clinked water bottles together.

"No tan lines."

Finally, muddied and quiet, we emerged onto the bush road within a few yards of the previous four teams and within a yard of a bright yellow flag.

The sun broke out as we did so. It was a welcome relief from the dreariness of tramping through the sodden bush.

Ben and two other crew leaders lounged in their trucks, windows rolled down. The rest sat on the road munching on apples and granola bars.

"What is this," Ben joked good-naturedly, "did all you guys play follow the leader? You came out within fifty feet of each other."

"Not everyone," I offered, jumping up onto the tailgate of Ben's truck.

"Jimmy and Debbie are on an Arctic expedition," Sandy grumbled through a mouthful of granola.

Everyone but Ben laughed. Ben checked his watch, his face dark with concern.

The rest of us stripped off our rain gear and jackets and lounged in the sun.

"We'll give them half an hour and then we go look for them," Ben said.

"Can't we just leave them to eat each other. You know, survival of the fittest, like the Donner party," Colin suggested.

We all burst into a fit of giggles until we realized that Colin was serious.

"Colin, that is very cruel," Colin's partner, Ocean, gently criticized, her voice as soft as a kitten's mew.

Ocean was as gentle as her name suggested. She was a tall willowy girl with long auburn hair and a gentle compassion for everyone and every living thing.

It was an odd match and I pitied Ocean. Colin was a jerk.

"It's not cruel. If we ever get lost for more than a couple of days, I will eat you if I have to."

There was stunned silence.

Ben glanced sideways at Colin. He then studied each and every one of us. For a moment, Ben's and my eyes met. I sensed that we were thinking the same thing. Find Ocean another partner and delegate Colin to the office!

A series of oaths came from the surrounding bush.

Debbie and Jimmy stumbled out of an alder thicket, worn and frazzled. Sweat dripped from their brows. Neither spoke.

Sandy started to hum 'Born to be Wild' by Steppenwolf.

I elbowed her in the side. She grinned, her face beaming with happiness.

It hit me like a runaway freight train! Sandy really did have the hots for Jimmy. Duh! I got it. I finally got it!

"Alright everyone, nap time is over. In the trucks," Ben commanded. "We still have a full afternoon ahead of us. We're heading back to the park and everyone will get a chance to drive a Terrajet. Those of you who aren't on fly-ins will probably, at some point in the summer, have to use one."

"Woohoo!" one cruiser cried out.

"Awesome!" Deb cheered up.

"They aren't bumper cars," Ben growled.

"But I want everyone to have a ride for their money!"

"Only your partner," Ben admonished Debbie.

Deb sulked.

"Not happening," Jimmy growled.

"God help you, Jimmy," someone joked.

"I've driven one. They are a riot. Terrajets can go through almost anything, unfortunately, they do sink," Sandy said.

"And how would you know that?" I asked her, alarmed.

Sandy shrugged guiltily.

"Sandy. Laura. You're in the truck with me," Ben ordered.

My heart sank.

Three trucks sped down the highway; Ben was in the lead. As Ben drove a single cab 4x4, there was only room for the three of us in the cab. Sandy sat between Ben and I, her legs splayed open because of the 4x4 stick shift on the floor.

"Sandy," Ben said her name as if it was a divine prayer.

"Ben," Sandy replied.

"Sandy. No James Bonding this year."

"For you, anything."

"Glad to hear it."

I pulled a squished ham sandwich from my coat pocket and unwrapped the waxed paper.

"You may not want to eat that," Ben offered.

"Why?" I asked through a mouthful of sandwich.

Sandy and Ben started to giggle.

"Terrajets are bush buggies. They have a stick shift, a brake pedal and a steering wheel. They are noisy as hell and bounce you around like a ball in a spinning barrel," Ben informed me.

"The brakes take a while to work so when you need to stop, don't aim it at the lake," Sandy added sagely.

"There aren't any seat belts and they tip easily too," Ben nudged Sandy good-naturedly.

"Oh," was all I could think of to say as I tucked my sandwich away.

I looked in the rearview mirror. The others were gobbling down their sandwiches.

"Should we stop to tell the others?"

"Nope," Ben and Sandy said together.

"To quote Ocean, 'that's really mean'."

Ben and Sandy laughed.

When we arrived at the gravel pits, I noticed that the walls were at about a sixty degree angle. A gravel track wound its way through the middle of the pit in a casual loop. The walls of the pit were overgrown with hardy grasses and shoots of willow. A few scattered boulders nestled against the sides of the road and at the base of the walls.

It was a great test site for Terrajetting, unless Mario Andretti was your driver.

Once again, crews went in order of their number. Sandy and I were last.

Crew after crew spilled their beans over the side of the Terrajet. I appreciated Sandy for warning me. Ben, I wasn't so sure about anymore, given the delight he and Sandy were sharing over the contents of each spew.

Colin and a white-faced Ocean careened towards us. He was driving. We dived out of the way as he barreled right through the middle of our group. I swear he was trying to hit us on purpose. He missed Sean by a few inches. More than ever, I felt Ben should relieve him of duties.

"Crew six, you're up," Ben yelled at Sandy and I after giving Colin a stern warning.

"Did you move the testing site to the pit just for me?" Sandy asked Ben, batting her eyes.

"Yep, just for you, darling."

"I'm driving," I said, stepping into the driver's seat, but Sandy hip-checked me out of the way.

"Not!"

"Sannnndy!"

I nervously got into the passenger seat. I looked around for a seatbelt and then remembered that there weren't any.

"Lolllyyyy."

Sandy turned on the engine and gunned it up the main road, far away from prying eyes. The noise was deafening. I thought my kidneys were going to sprout out of my eye sockets as we bounced from side-to-side.

I glanced over my shoulder at the empty roadbed and stark grey stone walls of the pit.

I placed my feet against the dash and pressed my spine hard against the seat back to keep from being bounced right out of the machine as Sandy aimed for the steep slope. Sandy laughed like a madwoman beside me.

"Slow down, you're going too fast," I cried.

"Hang on, we're going to the top!"

"The top of what?" I screamed.

Sandy grinned and raced up the side of the pit.

"I don't want to go to the flipping top of the pit," I wailed, my stomach clenching. I was so glad I didn't eat that sandwich.

We made it three quarters of the way up and then the Terrajet's wheels started to spin on the loose shale.

Sandy whoo-hooed in delight. She clenched the steering wheel so hard that her knuckles turned white. She tried to turn the wheel, but it jumped out of her hands.

The Terrajet's engine screeched.

The buggy tipped sideways.

"Throw your weight towards me," Sandy hollered.

I grabbed hold of the top of the windshield and threw my shoulder against hers.

The buggy turned on two wheels, but didn't flip.

We careened down the pit's walls in a landslide of gravel and loose shale. We bounced off the roadbed, out-of-control, the right tire catching a boulder. Metal screamed in protest as we slammed to a sudden halt. I hurtled over the windshield and tucked myself into a ball in the air, knowing the landing was going to hurt. It did. I landed on the gravel with a thud and rolled to a stop.

'Lord, what kind of madwoman am I working with,' I wondered as I sat up. The ground whirled around me like I was on a merry-go-round.

I picked myself up, dusted off my jeans and jacket, and then checked myself all over for broken bones. I was still a bit dazed, but thankful that nothing appeared broken.

"Aw, heck," I muttered, examining the dented fender.

Sandy turned off the revving motor and looked over the damage.

"It's only a little dent."

"You are frigging crazy, you know that?"

"Yeah, but we're going to have a great time together," Sandy laughed, rubbing her ribcage. "Oh, man, I think I bruised my ribs."

"No kidding."

"You know we're getting paid for this? How cool is that?"

We worked together and pushed the Terrajet off the rock.

"Don't even go there," I said, getting into the driver's seat and starting up the engine.

"Spoilsport," she said.

I took off slowly, but by the time I got around the bend, I had the pedal to the floor. I forgot about the braking part and careened into a patch of alder on the far side of the road just before I hit Ben and the others.

Ben raced over to us and switched off the motor.

"Good God, what were you thinking, lady," he growled, ruing the day he put two girls together on a crew.

"Sorry, I forgot about the no braking part," I wheedled. And then it hit me and I grinned. I'd never been called a lady before.

"And we didn't put the dent in it either," Sandy said, defensively. "Just so you know."

"What dent?" Ben said angrily.

🌿 🌿 🌿

The weather turned from rain with a chance of flurries to Bermuda balmy overnight as it does quite often in the Canadian northern shield region.

The poplar, alder, oak and maple trees began to unravel their leaves. Cedar trees released buckets of yellow pollen.

The bears came out of hibernation. They were sleepy, grumpy and incredibly hungry. Unpredictable at the best of times, they were worst in the spring.

Bears whose dens were close to town or the provincial parks looked for easy pickings at local garbage dumps and campsite dumpster bins.

The northern black bear is not Whinnie The Pooh. Beautiful, yes, with sleek and glossy, almost blue-black silken coats and dark liquid eyes, but dangerous and contrary by nature.

The Great Lakes region gets flooded with Great White Hunters, all itching to shoot a spring bear at this time of year. Some of these fine hunters sit drinking beer in their trucks at the dump until the first poor sod of a bear wanders in front of their truck, and then BLAM, the bear is dead, skinned, its carcass left to rot. Other hunters are in it for the hunt and meat. These men and women slink through the bush until they find their target, leaving nothing behind when done, neither pelt, bone nor meat wasted.

These days, eco-tourism is the buzz word, and the camera has replaced the rifle as the weapon of choice.

On this fine sunny spring morning at six a.m., we awoke to a series of high pitched shrieks. It sounded like the world was coming to an end.

All twelve of us dashed out of the bunk house in various stages of dress, from night robes to shorts, lace camisoles to jeans.

Ben raced out of the senior officer's quarters zipping up the fly on his jeans, shirt undone.

Even through sleep encrusted eyes and a befuddled brain, I smiled. The man was hot!

Yeah, I was engaged at the time, but a woman can look, can't she?

Ben took in the kitchen's shredded screen door and the shattered boards of the inside wooden door. The cook's keys were still dangling in the lock. "Get back everyone," he immediately cautioned.

Another shriek and a series of loud thumps came from inside the cook house. The thumps were followed by a bawling noise, much like a calf makes when it is afraid and can't find its mother.

"Seriously back-off, guys!" Ben commanded.

All at once, the half-closed door banged open and a bear raced out of the cook house. It slipped and slid on buttery, maple sugar covered paws. Spots of white flour speckled its black coat. Syrup dripped from its lips.

Our middle-aged, slightly over-weight, five foot three inch German cook whose pastry was fit for the gods and whose stew was more scrumptious than my grandmother's, raced out after the bear. She beat it across the ass with a straw broom, screaming in German at the top of her lungs!

The young bear, no more than two years-old as he was small, weighing maybe 300 lbs, bolted across the parking lot, heading for the forest as fast as he could.

Ingrid shook a fist at the bear, and then looped the broom under one arm. She noticed us standing there, mouths gaping open in shock, and smoothed back her hair.

"No hot breakfast this morning, ya? D'ere is Quaker Oats, maybe."

"Uh, can we help you clean up in there, Ingrid?" Ben asked, a little shaken.

"Ya, that would be nice. You kids can clean the dining room, I v'ill look after the kitchen and put some coffee on," Ingrid agreed before disappearing back into the cook house.

We looked at each other innocently and then began to laugh.

"Ya," Ben mimicked Ingrid. "Get dressed boys and girls, we've got some cleaning up to do."

Several minutes later, we were back to help Ingrid clean up. The cook house was trashed.

Twenty pounds of flour was scattered across the dining room and kitchen like fresh snow. It layered the counters and table tops like Mother Nature had waved her magic wand and turned the cook house into a white wonderland. It seemed fitting that an empty Robin Hood bag sat open in the middle of the main aisle given what fun the bear had obviously had. Butter and margarine clumped with the flour into cookie dough mounds in places. Maple syrup dripped off of counters and congealed into sticky piles of goo.

A garbage can was over turned in the corner. Used coffee grounds and left-over spaghetti from last night's dinner created orange and black webs across the floor. It reminded me of a Dali painting.

Two large maple syrup paw prints skidded across the floor by the far wall. Several pieces of straw broom were intermingled with the prints. There was one giant print about six feet off the floor which meant the bear had to have been standing on its hind feet.

Sandy and I exchanged a look.

It didn't take a genius to figure it out. Ingrid had cornered the bear. She beat it with the broom while it stood bawling on its hind feet.

"Remind me to keep my mouth shut if Ingrid ever makes something that I don't like," Sandy whispered.

"Ditto," I said.

The pit in my stomach widened, not all of it from hunger.

Chapter Three

In Peter Pan's Shoes

We finished training camp and were lucky to have a couple of days off. Ben staggered the crews start dates so that some of them went to work right away and some had four days off before heading into the bush by truck or float plane. Sandy and I stayed at the park and just hung out, reading and boating on the lake. We had also gone for a day trip shopping into Elliot Lake with Debbie and Ocean.

By nine a.m. on the day we were to begin our ten days in and four days out routine, we were packed and ready to head out on our first tour of duty as a team.

Sandy and I were a road crew. Jimmy and Deb were on a fly-in as were Colin and Ocean. These were the people that I had gotten to know the best during training camp; although, I'm not sure anyone was truly able to get a handle on Colin.

Sandy and I were issued a 4X4 and a square stern sixteen foot aluminum canoe with a small 2 HP motor as well as a two man tent, a variety of camping gear, and four days of fresh meat and milk, the rest being dry rations. There was no popping down to the store for a loaf of bread or jug of milk during our ten days in the bush. You ate the fresh meat in the following order: chicken, pork, and then beef. After that, it was canned goods or catch your own fish.

I still can't stomach tinned ham, corned beef or spam to this day.

We checked and double checked our equipment and supplies before heading out. We drove north on highway 639 using a county map to navigate by. Our camp site was circled in black magic marker.

Sandy drove. I navigated.

I know what you're thinking! You let her drive? Again? Yes, I did. It made her happy and a happy partner is easier to live with when you have nowhere else to go and no one else to talk to for ten days.

Anyone who has ever tried to find their way with a county map knows that the maps are grossly inadequate. An hour and a half of driving behind us, we pulled over by a stream and came to the conclusion: we had no idea which secondary road we were actually on after having left the highway behind us. One gravel road looked like every other.

Like any smart woman would do, we decided to wiggle our toes in the stream and contemplate life for awhile. Eventually we knew a logging truck would come by and we could discreetly ask directions. We were damned if we were going to openly admit to anyone that we were lost.

About thirty minutes later, our savior arrived.

The truck driver was about sixty with a day's growth of whiskers on his chin, a worn out baseball camp with illegible lettering on the bill, and a ready smile. With a screech of rusty brakes, he stopped his truck on the bridge and turned off the noisy diesel engine.

"Catch any fish with those wigglers," he asked, leaning out the open window.

"No. We just stopped for lunch and a coffee break," Sandy answered.

"I see."

I felt myself blushing.

"Where's your coffee," he asked politely.

"Done. We've been sitting here awhile," Sandy responded.

"Hmmm. Where you two lovely ladies headed?"

"Just down the road a piece."

"Not much down this road except a skidder and a pile of logs."

"We're looking for a private camping spot so 'not much' sounds about right."

"Is that a fact? Well, there's no place to camp down there. The nearest lake is over on the next road back from where you turned off the highway." The trucker's eyes sparkled. "But I expect you know that, what with you gals working for the Department and all."

"We do, but we thought this road might lead around to the backside of the lake," Sandy lied.

"How do you know that we work for the Department," I asked, curious.

"Cause that's what the sticker says on your canoe and truck, unless you stole them."

The trucker grinned. We were done for. He knew we were lost and knew where we worked!

"Uh, so which lake is that exactly?"

"Long Lake, honey, there ain't no other one close by."

Sandy shot me an evil look before saying, "Thanks."

"We'll maybe give that one a try," I said softly.

He started up the engine.

We thanked him again.

In the driver's side mirror, I saw him wink at me as he rumbled off, our lungs choking on the diesel fumes and dust he left in his wake.

"At least he didn't make us admit it," I said.

"Doesn't matter," Sandy replied, heading back to the truck. She pulled out the county map. "Long Lake is where we're supposed to be."

"You think Ben is going to know?"

"Oh, yeah," Sandy said, folding the map into quarters. "Everyone will."

"Damn!"

"In spades."

Long Lake was a pretty lake with steep sides and a rocky bottom. Red and white pines lined the ridges. Willow and aspen caressed the water with feathery hands in the gullies. The lake was about one hundred feet deep at its deepest point and abundant in lake trout. Speckled trout filled the tributary streams that ran into it.

One end of the lake was marshy. Water striders floated lazily in the rush filled pools at the southern tip.

Loons called out day and night with hardly an intrusion

Whip-poor-wills haunted the morning the grey dawn and evening hours.

We pitched our tent in a clearing along the shore. Beavers had cleared most of the aspen out of the area so it was relatively open.

Hunters and fishermen had widened the camping area and cleared a track down to the lake, large enough for us to drive our big old Dodge down to the water. They had also fashioned a crude rock rimmed fire pit and a make-shift log table.

We were happy to be able to park the truck close to the tent. We were two girls alone off the beaten track. Security, even at this remote spot, was an issue. It was also nice to keep the food in the truck at night, away from critters.

It took the rest of the afternoon to set up our campsite. Once done, we made a fire and cooked up an early dinner of bbq chicken and rice, before heading out for a paddle. We wanted to get the lay of the lake before we went to work in the morning.

The water was as dark and calm as the surface of a mica mirror. The surrounding hills were reflected in the water. Gold and brilliant reds reflected off of the lake in a kaleidoscope of color as the sun dipped down below the horizon. The young white birch and silver aspen along the shore had an other-worldly aura.

Loons' poignant calls echoed in the twilight.

Small darting slivers of silver jumped at the water bugs that swirled in myriads along the surface. Even a minuscule splash sounded larger than ever in the stillness of the evening.

"Wow, this is incredible," I whispered to my partner.

"Yeah, I know. No lights. No traffic. Just the universe."

"I feel like an intruder. A stranger in a strange land and all that."

We sat in silence for a moment, letting the canoe drift lazily in circles. The water sloshed lightly against the canoe's aluminum hull.

"This is the best time of year," Sandy whispered back. "No flies. Man, when the black flies start, look out. We are going to need daily blood transfusions."

I laughed.

"This is going to take some getting used to," I admitted.

"Why?"

"It's so quiet. I mean…where's the sirens…the roar of the jetliners? I live near the airport in Ottawa."

Sandy chuckled.

"After a while, when you're out, you won't be able to sleep until you get back in."

"I guess," I said, unconvinced.

Sandy and I huddled into our vests as the sun slowly sank behind the hills and the creeping chill of evening sank over the wilderness. We picked up our paddles and paddled slowly back to camp.

With a soft thud, the canoe grounded itself on the patch of graveled earth that was going to be our home for the next week. We pulled the canoe up on shore and crept toward the tent with only starlight and a sliver of moon to guide us.

The fire on which we had cooked our dinner was nothing but a few charred embers. We had made sure we doused it well before leaving camp.

We chatted quietly as we tucked everything away for the night. Coolers went into the truck cab. Canned goods were left in the box in the truck bed.

"Want to get up with the birds? We can work early and maybe take a couple hours off in the afternoon?"

"Works for me," I agreed.

"We might even get to see some moose. They usually come down to the lake in the early mornings for food and water. Deer too."

"That would be cool. I've never seen a moose in living color."

The radio inside the tent crackled.

"Blind River Two, checking in," blared from the speaker. "Over and out."

"I didn't realize I left the radio on," Sandy said, worried. "Hope we don't lose the battery before we're done."

"Yeah," I replied, equally worried.

We slipped inside the tent and listened as each of the crews checked into base.

"You're the crew boss, officially," Sandy nodded towards the mike, "so you better check in."

"Do I sense a little bit of Chicken Little coming on?"

"Noooo," she wheedled.

"Fine," I said, picking up the mike. "Blind River Six, checking in."

"Over," Sandy reminded me.

"Over."

"Blind River Six, this is base. Over," Ben's baritone voice crackled over the speaker. "I hear you had a nice lunch on the bridge. Over."

Sandy and I burst into a fit of giggles. Sandy motioned for the mike.

"We sure did, base. Met us a mighty nice fellow too. Told us the best place to camp. Over."

Sandy and I roared with laughter.

"Glad you liked my dad. Over," came the reply.

We stopped laughing.

"The early bird catches the worms. Off to bed now, base. Over," Sandy said, with a helpless shrug.

"Nighty-night, ladies. You too, everyone. Over."

"Night, John-boy. Over," crackled Jimmy.

"Night, Jim-Bob. Over," said Colin.

"Night, Mary-Ellen. Over."

The Waltons was a favorite TV show back then.

Soon the radio went silent.

Sandy and I glanced at each other and fell over laughing. We were off to a great start.

I didn't get much sleep. The rustling in the bush beside the tent all night had me wondering what was making the noise: a mouse, rat, skunk, porcupine? The list was endless.

And then there were the crickets and the frogs. Oh my lord they were loud! I thought it was going to be quiet in the bush! Nope!

The night's orchestra was beautiful and mystical, but in other ways, it was downright annoying.

I finally drifted off to sleep, but woke with a start. Something nuzzled my foot. My sleeping bag touched the wall of the tent. It was only a two man tent and we had air mattresses, sleeping bags, and our back packs inside it. I was startled awake by a blackness blacker than the surrounding night a few inches from my toes.

I was terrified, too afraid to wake Sandy thinking that she might yell and scare whatever it was into attacking. My gut told me that it was a bear. I envisioned giant claws tearing my skin and the thin nylon fabric of the tent into confetti.

My heart pounded like a kettle drum. The shadow moved away moments later.

It was an hour before I was able to fall back into a restless sleep.

I woke up with a groan. I was groggy and tired. My air mattress had deflated in the night and my back ached from sleeping on the ground.

"Coffee or tea," I asked Sandy as I climbed out of my sleeping bag.

"Coffee." She yawned. "Call me when it's done."

She rolled over and went back to sleep.

I dressed quickly into jeans and lumberjack shirt, tugged on my Grebb steel toed boots and unzipped the fly of the tent.

It was a beautiful morning, I thought, as I strolled into the bush with a roll of toilet paper and went about my morning business.

The only sound was my heavy breathing. No frogs. No crickets. No birds trilling. It was like the world had drawn in its breath and not yet let it out.

Every step or move I made seemed incredibly loud to me.

There wasn't a breath of wind. The sun hadn't risen over the hills yet. The day was steel-dust grey, but filled with promise.

Thin tendrils of fog drifted over the lake.

The ground was wet with dew.

It was the false dawn, that eerie time of morning when native peoples believe that lost souls walk the earth.

I stretched and yawned.

This day was going to separate me from the girl that I was. This moment was the reason that I chose to go into forestry in the first place. I wanted to make a difference and, however small, this was the time and the place to make it.

I was in the bush. I was miles from civilization. I was about to walk into the bush this morning with nothing but a compass to guide me and a partner to watch my back. I trained for two years for this!

I was as nuts as my partner, I suddenly realized and let out a soft chuckle.

I lifted the lid on the Coleman stove and read the directions. Most of the words had been rubbed or burnt off from the previous year's users.

I tried to remember what Ben had said to do and not to do in our session last week. My stomach growled at the effort.

What the heck, I thought, it can't be too hard.

I pumped the little handled on the side, turned on the gas valve, stood back three feet and threw a match at it. With a whoosh, the flame leapt up into the air. To my horror, it got bigger. I smelled burning paint.

"What's that burning," came the muffled words from inside the tent.

"Uh, the stove," I said.

"The what?"

The paint on the stove bubbled and burned.

"The stove's on fire!"

"The stove's on what?"

"Fire!"

Sandy careened out of the tent, clad only in long johns.

"Close the lid and turn down the gas valve," she screamed.

I grabbed a stick and used it to shut the lid. Flames leapt out from either side of the closed lid.

"Now what?"

"Turn down the gas valve."

"You turn it down."

"You started it!"

"I'm not going near it until the flames die down."

We stood there for several minutes watching the flames continue to burn the last of the paint from the stove. Finally, they began to subside.

I used the stick to lift the lid back up and placed a pot of water on the stove.

"Coffee's on," I chirped.

"You jerk," Sandy said, rubbing her eyes. She went inside the tent and grabbed her boots.

Sandy was quite the sight when she emerged from the tent. Disheveled hair. Red-rimmed eyes. White long johns. Steel-toed boots.

I started laughing. Sandy joined in. We threw our arms around each other and hugged deeply.

"Next time, you make the coffee."

After a filling breakfast of crispy bacon, scrambled eggs, and hot coffee, we made ourselves a couple of ham sandwiches and packed up for the day. We hitched our work packs over our shoulders and headed down to the canoe.

We tossed in the oars and life jackets, and then fired up the 2 HP motor. It coughed into life with a belch of black smoke and a dull roar which shattered the peacefulness of the morning.

As we putted down the lake, we remained silent. The beauty before us was breathtaking.

The sun still hadn't risen yet. The lake was dark and mysterious. The water flat and as black as the night. Mist drifted by us in clouds of gossamer threads, hovering a few inches above the water line, dislodged only by the ripples of our passing.

"This is it," I said in a hushed tone after examining our map and coordinates. We were at the southern tip of the lake.

We tied the canoe to a large black spruce standing near the water's edge. Sandy tied orange flagging tape around the tree.

I took out my compass and took a bearing based on the coordinates written out for us on our map. I then examined the aerial photo.

The tree we flagged was visible on the map. It was tall for this area, about sixty feet and leaned over the water at a thirty degree angle.

"By the aerial photo, I can just make out this tree so we'll use it as a tie-point and traverse in to the first stand from here."

I carried the work pack with all our gear in it and Sandy carried the lunch pack with sandwiches, granola bars, water bottles and a couple of cans of Orange Crush.

'We have four lines to run and then about a one mile hike out. These plots don't loop so it's straight in and straight out," she said, going over the morning's plots. "Easy-peasy."

"Right-O," I said, and then smiled.

I beamed at Sandy's confidence in her compass abilities.

We walked into the bush, branches scraping against our bright yellow hard hats. The hard hats, while hot, kept branches out of our eyes and helped us spot each other in the bush if we were separated. They weren't just for safety. We also sprayed the tops with bug spray when necessary. It was a lot better than having it on our skin. Muscol, like Coke, can eat the paint off of a car.

About two hundred meters in, we started our cruise line.

A cruise line consists of a two hundred meter line with ten stops, twenty meters apart. A prism sweep is done with an age and height reading taken of the dominant tree species at three different intervals.

The compassman tallies the readings by making a series of dots in allocated squares on a small computer sheet (now done on a handheld computer). Health of trees including presence of disease or insect infestations is also recorded.

The cruiser reads out the trees by species as she pivots around a stationary point, keeping the prism aligned.

The prism, read correctly, measures trees over a certain basal area enabling stand density, stocking and volume to be calculated. The location of the plot is printed on a basal map as well as a stand number and the various data that has been collected. In this way, anyone interested in the area can find out what tree species are present as well as their age, height, stocking and timber value. It is useful for determining when, and if, the area is suitable for logging.

After doing one line, another bearing is taken and the next distance paced off until we arrive at the next stand to be cruised.

The stands are plotted by someone in the office. The cruisers locate the place to start from and do all the leg work.

Basal maps provide bearings and distances. It is easy to fall into the pattern of not checking the given data. North is always straight up on aerial photos and one can easily check compass bearings to plots keeping that in mind. The person plotting the maps is not always correct and it is the cruisers' responsibility not to get lost because of their or our sloppiness.

"How long do you think it will take us, Sandy?"

"I'm thinking four hours. It's pretty rough going."

"Me too. We have cliffs we're going to have to scale coming up."

We tramped on, completing the second line. We came upon a rocky escarpment. We estimated that as the crow flies one meter horizontally equaled triple that on our climb up the embankment.

The going was treacherous. There was frost at the higher elevations making the rocks slippery and dangerous. One wrong foot hold meant broken bones on the rocks below.

After climbing the cliff, the going got better. We alternated between bramble covered slopes and clearer debris littered plateaus. The stands of trees switched between vibrant Jack pine and closely packed disease ridden Balsam fir. The pine stands were easier walking. We had to crawl on hands and knees under the firs or walk doubled-over when we were lucky. It was exhausting and sweaty work.

Logged areas were the absolute worst. Poplar and birch saplings grew over the decaying stumps of rotting slash piles. It was slow and grueling walking.

This day, like every other, we cruised all different types of ground from high ridgelines to bracken filled, swampy sloughs.

By the time the sun reached its zenith, we were at our limit. We wanted to stop, but there was nowhere nice to take a break so we continued on until we came upon the grassy edges of a slowly meandering stream. It broadened into a deep pool just to the right. The water was the color of outhouse offal.

A long forgotten bank of tightly packed sticks lay strung out from one side of the pond to the other, indicating a vacant beaver lodge lay on the other side.

Water streamed over the decaying pile unhampered.

We had a choice to make: slog through the pond and spend the day soaking wet or walk over the beaver dam, staying dry but risking the chance of a broken ankle or leg? It was a tough call.

"Looks like a good place to cross."

"I don't know, Sandy. Those sticks look pretty slippery, and old."

"Well, we can go around, but that will add another half hour to our day. The other choice is to swim the slough."

"Think it's that deep?"

"Bottom is silt. Even if the water isn't that deep, you could sink another two in the muck," Sandy added.

"Alright, let's give the dam a go," I agreed, not entirely convinced.

We had already been working for four hours without a break so anything that would make the day shorter was worth a try. Our first estimate of time to complete our lines was way off.

We rolled up our pant bottoms and spread our arms out for balance. We carefully picked our way across the stick bridge, wary of the many holes beneath our boots.

Sandy reached the bank first. I lost my balance, but managed to correct it at the last minute and jumped to safety, landing in a noxious sludge on the far side.

Sandy was kind enough not to mention my mud-dripping sulfur smelling boot.

We stopped for lunch after that. We were weary and hungry. The ham sandwich was hot and limp, but tasted marvelous as did the Orange Crush chaser. I needed a good sugar rush to finish the day.

We slogged on. One more line to do.

After wading through a soggy, matted area filled with springy club mosses and sharp brambles, we emerged on a stand of young Jack pine. The walking was easer and we finished the cruise in record time.

We studied the aerial photo to determine the best way back to the canoe.

Sandy took a bearing, hoping it would cut off the worst of our day so far and bring us back to the newer beaver dam that we had crossed earlier that morning.

We struck out and in an hour and a half crossed our own tracks.

"Not bad, eh," Sandy said, patting herself on the back with one hand.

"Yeah, but let's boogie. We still have another half hour ahead of us and I want to maybe go for a swim."

"Seriously? Do you know how cold the lake is? It's only May."

"Great for the heart!"

"So is all this walking."

I laughed. Sandy walked ahead of me.

"Sandy?" I said quietly.

She turned to face me.

"Wrong way."

"Oh," she replied sheepishly and reviewed her compass bearing.

I realized that it was that easy! You're tired. You stop for a minute and think you know where you're going and then 'Bam' you're lost.

We hadn't gone ten paces following our tracks back to the lake when Sandy let out a low whistle.

"What?"

I didn't want to stop for Sandy to pull out her camera and snap off yet another roll of film. My bedroll was calling.

"Look down."

I looked down and paled. Embedded over top of our tracks was a hug paw print. The depth of the print far exceeded either of our own. Slightly pigeon-toed, the prints had deep indentations at the ends indicating a colossal bear with equally colossal claws. Sandy spread her hand inside the print. Her hand didn't touch the outer boundaries.

"He's a big one and he's tracking us!"

"I wonder how close he is," I whispered.

"Jeez, I'd like to see him."

"Are you nuts? This mother's huge. I'm not going after him just to take a look and see how big."

"Yeah, but what a story we'd have to tell when we get back to camp. No one will believe us. I've never seen tracks this big before."

"If you want to live to see any more of these, then we need to boogey on out of here. Come on, let's put on our Peter Pan shoes and fly out-a here."

"Let me at least get a picture of the tracks first," she said, slipping her pack off her shoulder.

I was choked. Sandy was serious. Sometimes, I thought she had a death wish.

A deep throaty chuffing sound came from the bush behind us.

Sandy slowly stood up.

The sound came again followed by a low hog-like grunt.

We turned slowly and then froze.

Standing not thirty paces from us was the grand-daddy of all black bears. He peered directly at us. He swayed from side-to-side, and then dropped down an all fours. His nose quivered. He could smell something but couldn't see us. Bears have great noses but bad eyesight.

We remained still, not a breath escaping our trembling lips.

I glanced at Sandy. She looked like she was going to toss her cookies. Sweat dripped down her face. I expected my own face mirrored hers.

My ribs and chest ached from trying to keep my breathing low and soft. My muscles were taught with tension. Sweat stung my eyes. My heart raced. I wanted to scream, but dared not make a sound.

The bear walked closer.

Sandy's eyes met mine. Fear reflected in them. Blood pulsed through the carotid artery in her neck.

The bear stopped not ten paces from us and started rummaging around in the bush. He lifted his head and sniffed, but then dropped down to continue to root for grubs.

Seconds ticked by.

The seconds turned into minutes.

Noisily, the bear turned and scrambled over a fallen log, heading deeper into the brush. He had forgotten what had made him so nervous earlier. His aged white snout snorted as he tore an old stump to pieces looking for dinner.

"Think we should run yet," I whispered, not sure if I could run. My legs had gone to sleep.

"Wait a couple more minutes, see if he wanders off," Sandy whispered back.

We watched the bear's black rump move to and fro as he demolished the stump. Finally, he moved off.

We slowly backed away, stopping every few feet to make sure he wasn't coming back. Finally, we couldn't hear or see him anymore.

We slunk away as quietly as a mouse under a sleeping cat's nose. In minutes, we were at the abandoned old beaver dam.

"Bridge or water?"

I looked fearfully over my shoulder.

"Water," I said. "We don't want to bring him back to investigate and we can't chance an injury."

"Agreed."

We stole wordlessly into the water like burglars, emerging dripping and mud-covered on the other side.

"God, that water smells like shit," Sandy said, wrinkling her nose

"At least it covers our scent," I agreed.

We took our bearings and quickly stumbled out of the bush, shoulders tense and ears straining. Nothing seemed to be following us. Our canoe never looked so good.

With a sense of relief, we motored back to camp.

We built a huge fire and then washed the mud off our clothes and boots in the lake, hanging them up by the fire to dry. We didn't talk much. Maybe it was exhaustion from the day or maybe it was the emptying out of adrenaline, I didn't know.

I looked at my partner at one point. She stared into the flames with red rimmed and hollow eyes, hands firmly clasped around a mug of steaming tea, shoulders drooping with weariness. Sandy's hair frizzed out the sides of a pink ball cap with 'Sudbury Dolls' stenciled on the front. I assumed that was a softball team as Sandy was from Sudbury.

I started to giggle.

Sandy looked up, startled out of her reverie.

"It's a good thing we're miles from civilization," I said.

"Why?"

"Look at us."

Sandy looked at me quizzically.

"We're two girls alone in the bush sitting by the fire wearing nothing but our flannel knickers."

Sandy examined herself. Boobs hung out of her sleeveless cream colored low-rider body shirt. She had pulled her flannel pj bottoms up over her knees so that her bare legs and feet were red from the heat of the fire. A wool blanket tossed over her shoulders warmed her back.

I was dressed in a Victorian era blue flowered floor length flannel nightgown with ribbons at the throat. My grandmother had given it to me before I left for college. My feet were also bare and toasted red by the fire.

We both started to laugh.

I laughed so hard I had to pee, and slipped into my still damp leather work boots before heading off into the bush.

I was surprised to see my breath streaming out in front of my face in a billowy cloud. A chill swept down my spine and I shivered.

The night around me was eerily quiet. One brave loon called out, its forlorn cry embracing the starry night. No frogs croaked. Not a cricket chirped. The temperature had fallen dramatically in a short time.

The flames of the fire reflected in the collection of freezing dew drops on the long sedge grasses around the camp. Frost was forming. I squatted and peed, finishing in record time.

I shuddered and returned to the warmth of the fire. We stayed up for some time, staring silently into the flames and watching the forest around us turn into a crystalline world of ice.

The next day was beautiful. The ground sparkled, the frost so thick it was like a thin layer of snow had fallen overnight. The chill stayed in the air all day, making our work quite pleasant.

The following days were more of the same. Sundrenched days and frosty mornings. Campfires flickering beneath a star-filled canopy. We were cautious in the bush, both wanting and not wanting to stumble upon Grandpa, the nickname for our bear. We never did see him again.

Chapter Four

On Fly-In's and A Boy With One Leg

Sandy and I sat quietly on the dock. Our gear was packed and waiting. A sturdy, red fiberglass canoe lay upturned on the dock beside us. Paddles and orange lifejackets were bundled up with rope beside it.

"I still can't believe that guy," I said.

"What guy?"

"Terry Fox."

"Terry who?"

"Seriously? Haven't you seen the news?"

"I have no idea who you're talking about," Sandy replied, completely stumped.

"He's the one-legged guy running across Canada raising funds for Cancer Research. He'll be in Sault Ste Marie around the middle of August if he keeps running at the pace he's at."

"Why would a one-legged guy want to do that?"

Seriously? She didn't get it?

"Man, he lost his leg to cancer. He was a star athlete and then ka-pow, he's in a wheel chair and his life has changed forever. He started running in St. John's, Newfoundland. He's in PEI now. Sandy, the guy's frigging amazing."

"I don't see what the big deal is," Sandy quipped. "Sorry I just don't."

I was fuming. What was wrong with her?

A tense silence fell between us.

I had lost my grandfather to cancer four years earlier.

"Is he cute?" Sandy asked.

"Does it matter?"

"To me it does," she said after a minute.

"Yeah, he's cute," I said.

"Then I'll watch the news when we get out."

A group of agitated merganser ducks buzzed over the water in squadron formation. The males were strikingly coloured with white bodies, dark green heads and a slender red bill. The females had rich cinnamon short crested heads. They squawked in protest as they whistled by.

A low murmur was heard. It increased in crescendo until the hills reverberated with the thunder of the DeHavilland Beaver's single turbo engine.

We ducked involuntarily as the large float planed zipped over the treetops behind us. The snub-nosed silver plane circled the lake, banking into the wind, and then landing in a flurry of white foam on the dancing waves. It chugged noisily up to the dock.

The pilot signaled us through his open window to back away until the propeller stopped revolving. He expertly guided the float plane alongside the dock. He turned at the last minute, cutting the engine, drifting softly into his final destination.

Sandy and I dashed forward after the float plane docked. We tossed mooring ropes to the pilot.

"Hi, my name's Keith," the pilot said amiably as he moored the plane.

Keith Harris was an average looking, slightly beer-bellied middle-aged bush pilot with a spider web of laugh lines around his grey eyes. His navy blue windbreaker fluttered in the stiff breeze off the lake.

"Sandy. My partner here is Laura," Sandy said, extending a hand.

He shook Sandy's hand and then gave me the once over.

I graduated from Sault College of Applied Arts & Technology with a Forest Technician's diploma in 1980, one of only ten girls amidst a sea of one hundred plus boys. When Keith looked me up and down, I felt like meat. Not one of the hundred young men I went to school with ever made me feel that way. It was so blatant that even Sandy noticed it.

"Hi," I said, pulling away from his too firm a grip.

Keith chuckled, bemused.

"Let's get the heavy stuff loaded first," Keith nodded towards the neat row of packs and boxes of food on the dock, "and then we'll tie the canoe onto the floats after that."

Sandy waved me away as I reached for the cooler full of fresh food. She hauled a pack over her shoulder and then handed it to Keith who disappeared with it into the back of the Beaver. It went that way until all that was left were the lifejackets and paddles and one small box of sundries. That was when I stepped in.

"It's going to be a little tight for you, little lady," Keith said to me, a boyish grin lighting up his face. "Hope you aren't claustrophobic?"

"I'll manage, I'm sure," I answered. "I don't need much room."

Sandy pitched the last small box at Keith's head. Keith snatched it out of the air before it fractured his nose. He laughed good-naturedly.

"You know where to fly us in to?" Sandy asked icily. "Our maps are in my backpack if not."

"I know where to put you down, don't worry. I've been to that lake a hundred times before. It's a lovely spot. Quite romantic with the right partner."

"I bet you've got some experience at that," I joked.

"Just a tad," Keith agreed, pinching his fingers together.

Sandy flushed for some unfathomable reason.

"How're we going to put the canoe on?" I wondered, looking at the plane's narrow skids and even smaller doorway.

"You two are going to flip it over onto the pontoons and I'll tie it to the skids so that it rests against the plane. You'll have to untie the plane first," Keith directed us.

We flipped the canoe on its side and then rolled it into place against the pontoons. We then stopped and looked questioningly at Keith standing sentinel inside the plane's doorway wondering how we were going to get in. The plane drifted a couple of feet from the dock. The canoe blocked our way to the pontoons.

"Give me your hand and leap over here. Climb over top of the canoe and I'll pull you in. That first step is a wet one though so be careful."

"You first," Sandy suggested.

"Thanks," I countered, warily.

She grinned.

I jumped, the toes of my boots skidding off the side of the canoe. Keith grabbed me by the arm and lifted me bodily into the plane. His hands lingered on my waist a little longer than needed as he helped me through the narrow doorway and into the small passenger seat.

Sandy leapt across the chasm effortlessly and vaulted over the canoe and into the plane without any help.

Keith whistled, impressed.

"Come on, kid, you're up front with me," he said to Sandy.

Sandy and Keith settled into the front seats.

The Beaver's windshield was scratched and covered with dead bugs. Keith checked in with the local tower and filed his flight plan before handing me and Sandy a set of headphones. He then started the engine. The blades whirled. He taxied across the water to the far end of the lake before turning the nose of the plane into the wind and revving up the engines. The noise was deafening, even with the headphones.

The engine whined in protest as Keith pushed the throttle forward and we shot across the lake like a rocket. The Beaver pulled up steeply, missing the treetops of the forest at the end of lake by inches. We slammed against our safety harnesses, gravity forcing our backs into the worn leather seats as we climbed high above the trees.

Thirty minutes later, we were circling a long narrow brownish hued lake. The lake flowed into and was part of one of the major rivers in the area. A dark blue spot on the horizon indicated the other larger lake into which we were to portage over the course of the nine days we were there.

The unnamed brackish lake didn't look like a romantic getaway to me. It looked like a mosquito and leech filled slough.

As we circled over the river, we noticed the water wasn't as high as it would normally be. We had had a fairly dry winter and spring so we were able to cruise this particular section in late May instead of July. The river should have been a raging torrent, but while the water did overflow its banks, the rapids we saw looked easily manageable.

We looped in for a final approach. A huge bull moose ran into the thick underbrush on the northern side of Brown Lake as we decided to call the smaller lake.

"Did you see that?" I asked into the headphones. "His rack was huge!"

"I've flown lots of hunters into this area. You'll probably see a lot more of those fellows while you're canoeing down the river," Keith answered.

Keith expertly landed the float plane with the lightest of thumps and taxied us down the lake.

"I bet we'll get some totally bodacious pictures," Sandy added. "I brought extra film."

"Be careful of those big boys," Keith advised them. "They are more dangerous in rutting season, but that is a few months away."

Keith steered the float plane over to a rocky escarpment on the south side of the lake. He cut the engine and the plane floated a few more yards before drifting to a stop. My kidneys were ever so grateful. The plane ride wasn't that smooth.

"You guys stay seated and I'll drop the canoe off the pontoons. You can lifejacket up and I'll hand you down your gear," he offered as he slipped out of the pilot's seat.

The plane rolled from side-to-side as he unfastened the ropes on the canoe and dropped it into the deep water.

Sandy and I put on our orange life jackets. They smelled of black mould and gasoline.

We cautiously climbed out onto the pontoons. I crouched down and held the canoe steady against the chop of the water against the pontoons while Sandy and Keith loaded out gear into the canoe.

"You should be quite comfortable camping on top of that escarpment," Keith said, pointing towards a rocky embankment.

"We do have a couple of plots at this end of the lake," Sandy acknowledged. "I remember seeing this outcropping on the aerial photos."

"That's cool. Maybe we can cruise them this afternoon after we set up camp, get a head start," I agreed.

"Well, good luck ladies. I'll see you in nine days. There is a wide patch of gravely beach on the south-western shore of the next lake over. You can't miss it. I'll pick you up off shore. I can't get too close because of the sand bars, but it is an easy spot to find."

"Okay. Thanks, Keith," I said, wanting to appear more cordial than I felt. Never anger the bush pilot whose job it was to come get you when your tour was over, my inner voice told me.

With Sandy at the bow of the canoe and me in the stern, we pushed ourselves off. I had more experience paddling white water than Sandy did so I took the 'driver's' seat in the rear.

The sun's reflection off of the water was so bright that we could barely see the shoreline. We grabbed our sunglasses out of our packs and then back-paddled away from the float plane. We waved to Keith.

Keith made his way back to the cockpit. He slid open the window and waved to us.

"Hope you brought your rain suits," he shouted. "Rain's coming. Enjoy the next couple of days of sun."

We waved again, and then paddled hard to give the float plane plenty of room to turn. It was obvious that Sandy wasn't much of a paddler as she flipped her paddle from side to side in the canoe, almost losing it at one point. I hadn't noticed before because we were using the 2 hp motor on the square stern canoe.

Keith laughed at us as he flipped the ignition switch. The propellers spun into life, the engine whining loudly as Keith turned the rudders and taxied back down the lake.

With mixed feelings of glee and trepidation, I watched the plane take off. Keith tipped the Beaver's wings in salute when he buzzed by us.

"You really think it's going to rain?"

"Dunno," Sandy said, the canoe scraping against rock.

I side-paddled the canoe until it was flush with the escarpment. Sandy stepped onto the rocks and held it steady for me to get out.

"I know if it rains hard, the river'll swell. We might get stuck here."

That was a sobering thought: stuck on the shores of a dirty-brown lake that I didn't even want to dip my toes in.

"Let's set up the tent, grab a quick bite to eat and rocket off those lines," Sandy suggested.

"Works for me," I agreed.

After a quick lunch, we pitched the tent. We hung the fresh food cooler in a tree high enough to hopefully be out of reach of bears and raccoons, and then piled the rest of the dry stuff away from the tent.

We launched the canoe and paddled down the lake, heading for our first cruise line. Our wake looked like a drunken sailor's in a hurricane. The lake wasn't as calm as when we landed. It began to get choppy as the afternoon wind whipped the surface into a myriad of white topped waves.

"Hey, Hiawatha, let's take heap big canoe over to heap big shore," Sandy joked nervously.

"Are you okay?"

"This lake is a lot bigger than Long Lake," she said. "I admit, I'm not a great paddler and the other canoe was larger plus it had a motor."

"Let's practice then," I said. "Dip your paddle in and pull it towards your body in a 'J' stroke."

Sandy did as told and the canoe straightened out.

"Don't keep switching the paddle from side-to-side. I will guide the canoe from the back."

We continued across the lake until we reached a smaller rock outcropping.

"We're going to pull in here. Now, you are going to use an 'L' stroke to help me turn the canoe. Pull the water into towards you with the paddle on the side towards shore and we'll coast in."

We maneuvered into shore expertly.

"Well done, Sandy," I said, not having the heart to tell her that I had done most of the work from the stern.

She grinned, pleased with herself.

"That wasn't so bad," she said.

"No, it wasn't."

We tied up our trusty steed and headed off into the bush.

We were lucky. Most of the stands we cruised that afternoon were pine and poplar. We were finished the plots by four-thirty and the day so far had proved uneventful.

The lake was calm when we paddled back to camp.

After a dinner of pork chops and beans, I settled down to read a book. I had already devoured Carrie and The Dead Zone by Stephen King and was looking forward to reading The Stand. Maybe not the best books to be reading when camping in the bush, but it helped me ignore the myriads of black flies that chowed down on my face and arms all day long.

I made myself comfortable by the small fire we had built on the bluff over-looking the lake. The evening was cool and a slight breeze ruffled the water. A thin tendril of smoke washed over me. The smoke kept some of the flies at bay and the smell of burning pine filled my nostrils. It was sweet and bitter at the same time. I loved it.

I heard a sudden splash and looked up. There weren't any loons on the lake which spoke volumes about the condition of the water. Sandy had jumped into the canoe and pushed off from shore with a resounding curse. The canoe bobbed up and down, creating splash after splash.

"Sandy, what on earth are you doing?" I called, somewhat alarmed.

"Practicing," she said, resolutely.

I was instantly terrified.

Sandy kneeled in the canoe, about a third of the way from the middle.

"Take some pictures of me. I want to send some to my mom. My camera's beside my pillow in the tent."

I rummaged in the tent until I found her 35 mm camera under her pillow. I wondered why she kept the camera there. Was she taking pictures of me where I slept or was she waiting for something more exciting to happen in the night? Did I really want to know?

When I returned, Sandy was still doing circles in the middle of the lake. I smiled crookedly and snapped a few pictures. The camera snapped and whirred.

"Sandy, you need to sit in the larger seat, the one you were in this morning, and face forward. That's how you control the canoe if you are alone," I shouted.

Sandy sidled her bottom over the center brace in the middle of the canoe and made her way to what is classified as the front of the canoe (the wider section). She turned around and faced forward. She grinned and waved the paddle at me.

She paddled ferociously…in circles.

I tried not to laugh, but didn't succeed very well. My Stephen King novel was forgotten.

"Twist the paddle as you finish your stroke," I advised her. "Guide the canoe gently in a straight line."

A real trooper, Sandy continued paddling until she was finally able to successfully complete a figure eight. She then glided into shore.

We laughed the evening away. The campfire crackled. We wondered how the other crews were fairing.

When we had returned to base camp, just before we flew off that day, we heard that things weren't going well for Ocean. Colin was a jerk. I still didn't understand why Ben hadn't pulled him. He was creepy. It wasn't something you could put your finger on exactly, but I knew that I didn't ever want to spend a day alone in the bush with him and ten days was out of the question.

We were exhausted and exhilarated when we climbed into our sleeping bags around ten o'clock that night. We slept well…for a time.

We awoke in the blackness to howling winds and driving rain. The nylon tent snapped and strained against the pummeling it was taking. Rivers of water flowed down the escarpment beneath our tent. Water seeped through the seams.

"You tied up the canoe when you were done, right?" I whispered in the dark.

"I think so," Sandy murmured.

"Great!"

I rolled over and listened to the wind. There was nothing to be done in the dark anyway.

Gradually the heavy rain subsided and dawn broke upon a dismal drizzly day. Our mood was as sullen as the weather.

Our gear was soaked. Our sleeping bags were dry thanks to sleeping on air mattresses and my flannel nightgown was dry, but everything else was wet. I tugged on my damp jeans and boots. It wasn't a great way to start off the day.

We fired up the Coleman stove and made ourselves a hearty breakfast of hot coffee, ham and eggs. We were relieved to see that the canoe, though partially submerged in the water, was perched on a rock, its yellow mooring rope fastened around a large rock. Sandy had tied it up after all.

"Think we can stay here another day to see if the storm blows over?" Sandy asked.

"No, we won't finish our lines. We have to move and I don't want to be on this escarpment if a thunder and lightning storm rolls in."

"Good point," Sandy agreed.

We donned our bright yellow rain slickers and hardhats, and then covered our food and packs with heavy duty glad garbage bags before pulling down the tent and loading everything into the canoe.

By the time we had paddled half-way across the lake, the drizzle receded. A heavy layer of mist hung over the lake like we were in an Amazon rainforest rather than northern Ontario. Along the shore dew drops dripped from the bushes and trees into the still waters of the lake.

We reached the headwaters of the river and began paddling downstream, the current working for us instead of against us.

Orange columbine flowers and white trilliums bloomed beneath the shrouded evergreens. Our bright yellow figures and red canoe clashed with the vibrant greens of the forest and dark green-blue of the river.

A brooding stillness hung heavy in the air. The occasional 'plop' of a misplaced paddle sounded odd amidst the sighing of the river and the quiet solitude of the misty day. It was broken only by the caw of a crow somewhere in the distance.

An hour later, we stopped and let the current drift the canoe in towards shore. Sandy pulled out her aerial maps and examined them.

"Just around this bend in the river, we should see a big stump sticking out of the water. There's a cruise line that starts about two hundreds yard due east of there."

"It would be nice to set up camp first, I'm dripping wet under this rain suit," I said, sweating profusely. I may as well have paddled in just a t-shirt. It would have been more comfortable.

"From the aerial photos and maps, we have three lines that we can do directly off this section of the river."

"The bush is too thick here to camp anyway," I whined. I cursed myself for not agreeing to take a down day and continue to camp on the escarpment like Sandy had suggested.

"There's a small clearing down the river. Hopefully, it isn't too swampy. We can camp there."

Sandy handed me the aerial photo. I looked it over. She was right; there was a clearing about a half hour's paddle down the river.

Sighing wearily, my arms and legs feeling as lethargic as the morning, I slipped the mooring line over the old stump after swinging the stern around parallel to the shore.

We pushed the canoe gently offshore to hang in the current on the mooring line after we removed our day packs. The last thing we needed was to lose the radio and all our food to a scavenger just one day into our tour.

Sandy and I took off our rain pants. They were too hot to wear cruising. Our hands were already wrinkled from the humidity and rain.

Before we set off, we took out a jar of mink oil and coated our boots with the waterproofing oil. We changed into dry wool socks and were ready to go.

Sandy took a reading and we started off.

The black spruce stand we traversed was dense. The ground squelched underfoot. Water pooled in places. We walked on the exposed roots of trees and stomped down small bushes, trying to keep our feet dry. It was treacherous going as we leap-frogged from root to root.

Within an hour, we were soaking wet and ill-tempered. It was a blessing to reach the end of the plot.

I took a bore sample of a tall spindly black spruce and then stood counting its rings. Sandy counted out twenty paces and used the sunto to take a reading of its height. We recorded our findings, the pencil tearing the limp paper.

"I don't know how much of this I can take today," I admitted. "What do you think?"

"If we do the next line, we could go set up camp and dry out. The other plot line is just a short distance up river from that clearing."

"That sounds good."

We started compassing out, leaping from root to root and bush to bush. It was wet, slick and slippery going.

I jumped over a deep pool, but slipped on a root. My foot and leg went straight down into the hollow between the root system, submerging my leg to the thigh. Try as I might, I couldn't move. I was stuck doing the splits, my right leg out horizontally and the other disappearing into a black cavern of water.

"Yo, Sandy, help me out here."

Sandy laughed. She pulled a cheap Kodak camera out of her pack and snapped a picture of me stranded there.

"Not cool."

"Yeah, it is," she said with mirth.

She put the camera away and reached down. I grabbed her hand and hauled myself upwards. My leg and foot came out of the hole with a loud slurp.

"You ok," she asked seriously. "Your ankle isn't sprained, is it?"

I tentatively put some weight on it and walked slowly in a circle. My ankle was sore, but not twisted.

"I'm fine."

I limped along, following Sandy back to where we had tied the canoe. We were relieved to see it.

As we paddled down the river to the next plot, the clouds started to dissipate. Bits of blue sky poked holes in a dark storm cloud on the horizon. The forest awakened from slumber. Birds trilled. Frogs croaked. Black flies by the millions zeroed in on us.

We were in a much better mood when we completed the second plot, but it was a dirty go. We were wrinkled from the wet and blood-soaked from a thousand black fly bites. It was our job. We knew what we had signed up for.

When a brisk wind picked up in late afternoon, we were thrilled. The black flies diminished and the sun came out.

The river sparkled. Phantom sunrays danced between the trees. Water droplets blazed with rainbows of colourful prisms in the boughs of the spruce and cedar trees hanging over the water. The air smelled fresh and vibrant.

We stuck to our plan and found that the clearing we saw on the aerial photo was suitable for camping. We pitched our tent in minutes.

We tried to find enough dry kindling to start a fire, but quickly gave up. We lit the Coleman lantern instead and used its warmth to dry our fingers and toes before changing into clean and clammy t-shirts and jeans. Steam rose off our packs; the heat from our bodies combining with the lantern's turned the tent into a sauna.

We chilled for the rest of the afternoon, content to stretch cat-like on our sleeping bags and catch up on some reading. I dived into The Stand. Sandy, to my amusement, was reading Agatha Christie's Ten Little Indians.

A stupendously evil Randall Flagg confronted the angelic Mother Abigail at her home as she quietly rocked on the porch, praying to God and Jesus for guidance. The scene gave me the creeps, lying as I was inside a darkening tent, the sounds of a gurgling river and a thousand whining black flies at my back.

I put down the novel.

"Tomorrow, we can stay camped here and do the last line up river and another three down. What do you think?"

"Sure," Sandy said absently, engrossed in her book.

I tried to read again, but was too freaked out. I closed the book's cover and closed my eyes. Sleep came in fitful starts all night long.

🦢 🦢 🦢

The next few days on the river proved uneventful. Sandy and I settled into a rhythm. We adjusted to days of arduous paddling and cruising. Our muscles grew taught and Sandy slimmed down substantially. I was turning into a lean, mean climbing machine with thighs like an NLF's linebacker.

The weather alternated between sun and clouds. It was pleasant, but the black flies continued to suck our blood at an alarming rate. The horseflies too had hatched, taking huge chunks out of exposed flesh whenever they saw the opportunity.

We had yet to see or hear the moose we had seen when we flew in. We were disappointed, but also glad that we hadn't come across any bears either.

We had three more days to go and were down to our last rations of Kraft dinner, pancake flour, powdered soup and beans.

The river narrowed under a dense growth of willows and alder. We knew we were going to have a tough day of portaging ahead of us and then we had to find and set up a new camp. It was going to be difficult to get all of the last plots in. I cursed the head stand survey manager in the office for giving us so many plots to do. It was inhuman.

We found a likely place to stop to begin our half-mile portage to the mouth of the river. The mouth of the river ended in the large lake where Keith was going to pick us up.

Sandy and I unloaded our gear. It was going to take three trips to cart it all back and forth to the new camp. We didn't have any fresh food to worry about so at least we didn't have to hang the cooler in a tree, but we did put all the open flour and milk powder into it to prevent a raccoon from getting at it while we were gone.

We donned the lifejackets and roped the paddles under the seats of the canoe with the mooring lines. We then hoisted the canoe up over our heads and started walking inland into a tall poplar stand, leaving all our food and belongings behind us, neatly stacked and wrapped with orange ribbons so we could find it again. We needed to keep the river on our left, following the water sounds until we got to the lake.

We marched along at a steady pace, two red-faced, lumberjack-shirted figures carrying a red canoe through a green forest. Every so often, one of us would ribbon a tree so we could find our way back easier.

It was hot work, but a cool breeze blew in from the north which made it a little more bearable.

When we started to climb a small knoll topped with scrubby jack pine, I signaled a stop. We needed a break.

"Where you off to?" Sandy asked, leaning against a tree and routing her last Snickers bar out of her pocket.

"Nature calls."

"Don't get lost."

"I won't."

I headed down the slope towards the river, out of Sandy's line of sight.

There wasn't a good downed tree to be seen. Usually, I liked to find a narrow log on which to perch my skinny behind and do my business. That way, at least I wouldn't wind up with pee or otherwise stained jeans when I was done. Sometimes, I wished I was a man.

I squatted down and did my business as fast as possible. The black flies were bad and my butt cheeks were getting bloodier and itchier by the minute from all the bites.

I found Sandy dozing, her back against a tree, her face dripping with blood from a thousand bites. I started laughing, suspecting my face and buttocks looked the same.

"Let's rock and roll, girlfriend," I said.

Sandy grinned and stood up.

We picked the canoe back up and started marching forward. I was in the front.

There was a rustling in the bushes and out popped a large skunk.

"Stop! Stop! Stop!" I cried out.

"What?"

"It's a skunk," I whispered, the large black male skunk fixing his beady eye upon me. His tail went up. It twitched, but he didn't swing his butt around towards me...yet!

"Back up slowly," I commanded.

We slowly backed away while the skunk continued to twitch his tail in warning. Soon, we felt we were a safe distance away and we quickly scooted sideways, giving him a wide berth, and then splat!

"Shit!" I growled, lifting up my left foot. I shook it from side to side, but my foul droppings stayed fixed upon the underside of my inner boot sole.

"What?"

"Exactly."

"Huh?"

"I just stepped in my own crap," I said, disgusted.

Sandy laughed heartily as she ran sideways so that we were facing down the embankment towards the river so as not to step in it too. We still carried the canoe over our heads horizontal to the slope,

"It's not funny."

"Yeah, it is," she said.

The skunk looked at us oddly and then bustled off into the bush looking for someplace quieter.

I realized after a moment that we had just driven a skunk away in horror and idly wondered how many people could put that down on their resume?

We finished our portage. I dunked my foot in the lake, trying to remove the last of the noxious stuff from its sole. I swear the smell was with me all day even though my boots were bush scraped and spotlessly clean by the time we had completed three trips hauling all our gear back and forth from where we left it to the new camping spot.

🦨 🦨 🦨

The wind gusted up to forty miles per hour on the last day of the fly-in. It whipped the storm clouds gathered in the east across the sky like racing brigantines in full sail.

We didn't know if it was too windy for Keith to land the Beaver. We hoped not. We were out of food and neither of us thought to bring fishing gear.

We packed up our belongings, but waited until the last minute to pack up the tent. The lake was choppy and steadily getting worse.

We radioed base camp.

"Blind River Six to Base. Over," I said into the buzzing radio, unsure if we were going to get any reception. The white noise that came across the airways didn't bode well, especially with what looked like a whopper of a storm brewing overhead.

"Blind River Six, this is Base. Over," came Ocean's unmistakable soft and pretty voice after a few attempts.

Sandy and my eyebrows raised in puzzlement and concern.

"Ocean is that you? Over," I asked.

"It is. Long story. Over."

"We're just calling to see if Keith is going to be able to make it in to pick us up? It's pretty windy here and a major storm front is rolling in. Over."

"Let me check. Be back in a jiffy. Over."

We waited, hands steepled in prayer. After what seemed like forever, the radio finally crackled into life.

"He'll be there within the hour. He's dropping Crew Three off now and will be taking off again shortly. Over."

"Thanks, Ocean. We'll see you when we get back. Over."

I hung up the mike.

"Wow. What do you think happened?" Sandy asked me, concerned.

"I don't know, but at least Ocean is safe. I never liked that guy."

"Yeah, he was weird. That part about him being ready to eat Ocean if they got stranded. That wasn't funny."

"And I don't think he was kidding," I concluded.

We packed up the tent in silence.

We loaded the canoe and waited forty minutes before pushing off from shore.

I was looking forward to a hot bath and a shampoo. We had sponge baths in the tent at night because the lake water was too cold to swim in yet. We stank like locker room gym bags that haven't been washed in a month.

The mouth of the river where we had camped was covered in sandbars. Because of the waves, it was hard to tell how deep the water was where the river met the shallows of the lake. We would ground the canoe if we weren't careful. The thought of being grounded with a loaded canoe in a wind storm didn't do much to calm our nerves. We paddled slowly out into open water with this in mind.

The float plane had to land and take off into the wind so we paddled south-west with the winds at our back, in order to make it easier to pack and go once the plane arrived.

We paddled hard. Spray hit us in the face. The wind whipped the lake into a frenzy. Water pooled at the bottom of the canoe, soaking our gear. The weight from the wet packs and gear dragged the canoe lower in the water until the gunwales were only five inches above the water line.

"Maybe we should head to shore and dump the water out of the canoe before we go much farther," Sandy said.

"Yeah, I think we're going to have to."

We paddled as hard as we could. Even with the help of the wind, we struggled to get across the lake, trying in vain to get to shore.

We heard the Beaver's engine's roar and saw Keith fly in low over the tree tops.

"Guess we don't have time."

The silver Beaver swept above our heads. A wind gust slapped the plane sideways. Keith straightened it out at the last minute and successfully landed. The planed bobbed up and down on the waves as Keith eased the float plane towards the canoe, a white frothy wake streaming out behind. He shut the propeller off about twenty yards from the canoe.

We used the weight of the canoe to stabilize ourselves in the water. The wind was getting stronger and stronger. We paddled for all we were worth knowing that we only had a short window to get out of there.

The waters of the lake darkened with the darkening sky. The ducks and loons had all fled to the safety of the shore. That wasn't a good sign.

My back and shoulders screamed with the effort needed to keep the canoe in a moderately straight line.

Sandy turned once to look at me. Her face was a mixture of both fear and exhilaration. I had a feeling that my own expression mirrored hers.

It was us against the elements.

Keith stood on the pontoons, the door open behind him, the plane rocking back and forth violently.

"Move it, girls!"

A strong gust of wind caught us on the starboard side, sending a huge wave over the gunwales of the canoe.

We capsized

My horse bucked me sideways into a barn door once. I hit the door about six feet up, my back and head connecting with the wood with a solid thump before I fell breathless and semi-conscious to the ground. The cold water was so fierce that my exposed skin felt like it had been sliced with razors. I couldn't breathe. I had to reach the plane.

I struggled to stay afloat, even with the life vest on. My clothed and boots were weighting me down.

I slipped out of my life jacket and let it float away, knowing I needed get out of my rain coat and heavy flannel shirt. Sandy's life jacket, yellow raincoat, and plaid jacket followed after mine.

I tried to untie my boots but my fingers were too numb to grab hold of the laces. I gave up, striving instead to catch the errant life jacket.

I got mine back on, swallowing a bucket full of lake water in the process.

71

Keith grabbed Sandy with a grapple hook as she flutter-kicked hard towards him. She hung onto the hook for dear life as he hauled her to safety.

I tried to follow suit, but the hook just slipped through my frozen hands.

"Turn over and back kick," he ordered.

I did as told.

Keith hooked the grapple into the back of my life jacket and hauled me up and onto the floats. My teeth chattered like one of those wind-up chattering denture toys that you can buy at the dollar store.

"Get inside and wrap yourselves with blankets. You're both hypothermic," Keith ordered.

"What about our gear?" I asked, crawling into the plane.

"I'll fish it out."

"Okay," I whispered.

"Thanks," Sandy said hoarsely.

"Just call me Galahad. I rescue damsels in distress all the time, but you two are the prettiest so far this year. If you too don't stop shivering soon, we're going to have to strip down and get naked to get warm."

"Not!" Sandy and I said in unison.

Keith laughed and then grapple hooked our packs, paddles and hardhats. The cooler was lost to the lake. He then grabbed hold of the canoe and singled handedly dragged it onto the pontoons.

We helped where we could, the activity getting our blood circulating.

Once everything was tied down and secured, Keith taxied the plane back down to the far end of the lake, facing into the wind, readying it for takeoff. He turned the heat on full blast.

"You gals going to make it or is it time to get naked?"

We both shot him a thumbs-up sign.

He grinned and cranked up the engine.

We bumped over the lake. The roar of the wind and the whine of the engine was deafening, even with the headsets on. We took off, the plane careening wildly from side-to-side in the ever stronger winds. We were jarred against the back of our seats time and time again, the metal groaning in protest, as we took the tops off of several trees.

The pontoon with the canoe attached caught the top of big spruce tree. We heard a loud thump and a screech.

"What was that?"

"Sounds like we may have torn open a pontoon," Keith's calm voice said over the microphone.

"Seriously?" I asked, aghast. How were we going to land? It was a float plane!

"Don't worry," he said with a wink over his shoulder.

Keith expertly flew us back to base camp.

As we came in for the landing, he tilted the wings slightly so that the weight of the plane would land on the good pontoon. We raced across the lake. The dock and nearby beach came into view.

Keith didn't slow down.

He kept the plane on an angle, skating over the waves sideways, the throttle pushed forward.

We hit the beach with a resounding thump and skidded a few yards up it.

"See," he said, turning off the motor. "Don't worry. Safe and sound."

After we swallowed the lump in our throats, we climbed out of the float plane and onto the sandy beach.

Keith wouldn't allow us to unload any of our gear.

"Safety, girls. Safety! You need to get checked out?"

"No, we're okay" we both said.

"Thanks for the lift," Sandy cheerfully added.

"Ditto."

"Anytime. I hope you get to fly with me again."

We hauled our gear up to the beach head and waited there for one of the park officers to come down and help us. No one raised an eyebrow about the Beaver on the beach.

"Think we need to fill out an incident report?" I asked Sandy.

"Why? We're okay. The Beaver's Keith's problem."

"I guess," I said, non-committed.

We rolled our eyes at each other and started to chuckle. It was a good day. We were still alive.

Once we were warm and stuffed full of hot coffee and cookies by the cook, we drove back to Blind River for our four days off and went in search of Ocean.

Chapter Five

Ocean, Eight-legged Freaks & Breakfast For Three

Sandy and I had returned to the office after our four days off and sought out Ocean immediately. We had just missed her by a couple of hours by the time we drove back to Blind River from the park. We spirited Ocean away from the radio room, wanting to hear what had happened between her and Colin.

The following is Ocean's account of the story.

Colin was getting less focused each day. He said we couldn't be too careful and thought that people were watching us all the time: in camp, in the bush, in the middle of the lake…everywhere. He was always reading these survival books too.

He was a real gentleman when it came to respecting each other's privacy. He always let me get dressed in the tent first. He changed outside a lot or I'd leave and go make a cup of tea while he was changing. I never felt personally threatened until one night after dinner he started mumbling and glancing sideways at me.

It was a real strange look, one that got my skin to crawling.

He started talking about how to properly skin and smoke meat and then followed that up with how pretty I was and that maybe we should get married on our days off. It got crazier and crazier by the minute.

He started flipping his knife over and over in his hands, again and again. I asked him to stop and he started screaming at me.

I was terrified. It was like Colin wasn't even there anymore.

I knew I was in trouble and grabbed the radio. He grabbed me by the ankle and I had to kick him away. He tackled me, but I hit him with the radio. When he let go, I ran into the bush.

It was getting dark so I ran until I found a thicket of wild roses. I crawled underneath the sharp brambles and hid there until I heard Colin race by, still screaming my name and yelling that he was going to 'fix me'!

When I couldn't hear him screaming anymore I called base camp for help. Ben had me switch channels. I told him what happened and he told me to stay put. They couldn't fly in to get me until morning.

It was the worst night of my life. I was so scared. It got really cold too and I was dressed in just a t-shirt and shorts. The bugs ate me alive. Colin came by twice, waving a flashlight back and forth, but he didn't see me in the bushes. I had to pee but was too scared to leave my hidey-hole in case he was hiding out there too. I wound up peeing in my pants. I prayed and prayed all night long.

In the morning, I heard a couple of helicopters. Ben and John, the Head of the Natural Resources office, came in one chopper and the police arrived in the other.

I waited and waited and waited until I heard Ben and John calling my name that it was okay to come out of hiding.

When I came out, the police had Colin handcuffed and were taking him away in full restraints. He was still screaming.

I must have been a sight, covered in blood from a thousand bug bites and wearing pee-stained shorts. My arms and knees were shredded from the thorns.

Ben and John waited for me to get cleaned up and then packed up all of our equipment and gear. They switched me to office duties until Labour Day. After some of you guys return to school, they are going to match me up with one of you so that I can finish off the summer cruising again.

Sandy and I were stunned. It was a sobering conversation.

However crazy Sandy could get sometimes, it was nothing like what Ocean had experienced. I was glad Ocean was okay and found a new respect for my partner that day. No matter our differences, I resolved to deal with them.

After chatting with Ocean, we checked in at the supply depot and signed out a new truck and a sixteen foot aluminum boat with a 15 hp motor. The boat's hull was banged up and scarred from hard use. The trailer it sat on wasn't much better.

Sandy and I completed our four-corner check of the truck and trailer from tires to engine oil. We double checked that the gas tanks were full: the spare and the main tank in the boat. Sandy checked the hitch and tightened up the safety chains on the trailer. Everything seemed okay to go.

We pulled out of the yard and headed up the highway. This time, I was driving.

We were on the highway to Elliot Lake doing the speed limit, 55 mph, when I heard a bang. The steering wheel jumped in my hands. The truck swerved, the trailer pulling us side-to-side. My heart was in my throat.

"Uh-oh," I said, hearing a loud scraping sound. I looked in the side mirror and saw sparks flying out the back of the truck. The trailer banged forward under the bumper.

"The trailer's popped off the hitch," Sandy said, looking in her side mirror too.

I let my foot off the gas, turned on our four-way emergency lights, and then coasted over to the shoulder.

I turned off the truck. We walked slowly around the back of the truck to check on the boat and the trailer. The trailer hitch was wedged tightly under the black steel bumper of the truck, held firmly in place by the safety chains. The lock pin was broken off the ball.

"Thank you, God," I crossed myself before turning to my partner, "and thank you, Sandy."

"Wow, we were lucky," Sandy said.

Sandy and I high-fived each other.

We breathed a sigh of relief and unfastened the safety chains. We huffed and puffed, straining our backs, until we were finally able to get the trailer hitch back on the ball.

Sandy checked our tool box and found a spare pin for the trailer hitch. She clipped it on, double checking to make sure it was tight.

We stood on the truck's bumper and jumped up and down. The trailer was firmly hitched.

We breathed a sigh of relief and continued on our way to a small provincial campground on the shores of Flack Lake located on the western side of Mississagi Provincial Park. This would be our base camp for the next ten days. We had several different plot lines to do on a couple of the smaller lakes nearby as well as six days of work on Flack Lake itself.

We pitched our tent and settled down for the night. It had been a long day already. We fixed ourselves a nice steak dinner and went for a swim before retiring to the tent.

The night was pitch black. The stars twinkled, but there was no moon.

I had a fitful night and tossed and turned. I couldn't help but think of poor Ocean and her night alone in the bush when suddenly I heard laughter, lots of laughter.

I realized that I was back at the beaver pond and idly wondered how I got there? Sandy and I were wading across it, but I got stuck. I tried to get out, but the more I struggled, the deeper I sank into the muddy bottom.

I yelled for help.

Sandy turned around to help me, but she became stuck too.

The ooze and filthy pond water were up to our necks now. I sucked in my breath and tried to slip my arms over my head, but they wouldn't budge. Sandy and I sank slowly below the waterline, her eyes mirroring my own terror.

I gasped for air, but there was none. Blackness descended. It was cold. I started to shake.

I screamed! Pond scum filled my mouth with silt and offal.

"It's alright, Laura. Wake up," Sandy said, shaking me awake. "It's only a nightmare."

I blinked, the flashlight beam brighter than a burning sun in my eyes. I struggled to sit up. My sleeping bag was wound tightly around me, binding my legs and arms to my sides.

Sandy helped me untangle myself.

"Oh, my god," I stammered.

"That was a doozey, kiddo," she said, rubbing a hand across my back.

I shook uncontrollably.

"It was awful. I was drowning in that old beaver pond we swam through when we were running away from the bear. I got stuck in the mud and you couldn't help me."

"No wonder you were screaming blue murder," she consoled me.

"Sorry," I apologized.

We settled back down into our sleeping bags. I felt more than a little foolish.

"Sandy?"

"Yes, Lolly?"

"Mind if I sleep with the flashlight on by my bed for a bit?"

"No, go ahead," she replied sleepily.

In a few minutes, I heard a soft purring. Sandy was fast asleep.

I played with the flashlight for awhile, making shadow birds and churches on the tent walls until Sandy's soft snores filled the tent. I found some solace in that and turned off the flashlight, but sleep never came.

🦢🦢🦢

June 16th heralded the start of a series of beautiful and hot summer days. Dew moistened the ground. Clear liquid jewels hung suspended on leaves and beaded spider webs. Morning birds sang songs of joy. Squirrels chirruped gleefully in the trees around camp.

I was dressed by five am and ready to go. Sandy still snored peacefully inside the tent so I left her alone.

I made a pot of coffee and sat quietly at a wooden table looking out at the lake.

I wore loose fitting coveralls over a string bikini. The coveralls were cooler and kept the mosquitoes off of me better than anything else. They also had lots of pockets. I loved having a granola bar and a pop close by when I felt dizzy from heat exhaustion. There was nothing like a double dipper sugar fix to make you feel better.

We kept a water cooler bottle in the boat for when we returned to it.

"Do I smell coffee?"

"Yes, my lady, you do."

"Are you going to bring me one?" Sandy asked innocently.

"Nope."

"But I saved you from your nightmare," she wheedled.

"Waking saved me from my nightmare."

"But I woke you up."

"Too bad."

I laughed, amused.

I heard a series of shuffling noises. I ignored them and fixed myself a bowl of porridge. I resumed studying the flat waters of the lake.

I stood up and stretched before making my way to the outhouse, a roll of toilet paper in hand. An outhouse was a luxury item in my world. The smell was noxious, but not nearly as bad as the unrelenting itch of a dozen mosquito bites on your butt cheeks.

When I returned, Sandy was sitting sleepily at the table with a steaming mug of coffee in her hand.

"Well, what're we going to do today, boss?"

"We'll take the boat out across the lake and do the lines over there," I pointed west. "If we make good time, we can catch some rays and check out the other lines we have to do this week."

"Sounds good to me."

"Hope that boat doesn't leak," I said, looking over at the boat and trailer. "It's pretty battered."

"Won't matter if we add a few of our own then."

I chuckled, knowing we probably would.

"Ready to pitter-patter?"

"Yep," Sandy agreed, cleaning out her coffee mug. She wolfed down a granola bar and then snatched the keys from the table. She signaled for me to guide her down the boat ramp.

I was impressed. She started out well, carefully maneuvering the trailer into a straight line, but at the last minute gunning the accelerator too hard and jack-knifing the trailer.

"Turn the wheels in the opposite direction to what you want the trailer to go," I yelled. I had backed up a few horse trailers and knew what I was doing. "And slow down."

She tried again and jack-knifed the trailer a second time.

"Look, pull forward and straighten up, and then back up more slowly."

Sandy was annoyed. She spun the tires going forward, slammed the truck into reverse and jack-knifed the trailer a third time.

"Okay, okay," I said, walking towards the truck.

"You think you can do better, have at it," she said, stomping away.

I slipped into the driver's seat, pulled forward and slowly inched the truck and trailer backwards. This time, I jack-knifed it. I hadn't calculated on how light the boat and trailer were compared to a horse trailer. The horse trailer was much easier.

We both stood there, hands on our hips, contemplating the situation. We were glad that we were the only ones camping in this remote campsite too.

"There's only one thing to do," Sandy said.

"And what's that?"

"Unhook the sucker and push it in."

We looked over all the scrapes, dings and dents on the boat's hull and trailer's fenders.

"Don't think we'll be the first," I added.

We smiled.

"Ready, partner," she asked, manning the winch.

"Let's get'er done."

Sandy released the catch on the winch. The handle raced crazily in faster and faster circles. The boat flew off the back of the trailer, the stern landing with a splash in the water. The motor flipped up and then down, landing on its blade, firmly wedged in the sand. The rest of the boat lay half on and half off of the trailer.

"Damn!"

"Now what?" Sandy asked, surveying the damage.

"I got this," I said, running to the winch. I let out the rest of the line, and then climbed into the truck.

"Out of the way, Sandy."

"What on earth are you doing" she yelled, backing away from the boat.

I pulled forward with a spinning of tires and spew of gravel. Sandy covered her ears. The boat scraped noisily along the worn rollers. It swung off the trailer, the bow ricocheting into the air, hovering there for a minute until it fell with a sickening thud onto the sandy beach.

I pulled forward, parking the truck and trailer farther up the boat ramp in case anyone else came that day.

"And you think I'm crazy," Sandy said as I walked towards her with a spring in my step.

We pushed the boat the rest of the way into the water and stood there for a moment, waiting to see if it was going to spring a leak or not. It didn't so we loaded up our packs and then hopped in.

Sandy took the stern and reefed on the Evinrude's power cord. After a few tries, the motor sputtered into life.

Sandy threw the gear into reverse and eased the boat away from the boat ramp, turning swiftly as she did so. She then threw the lever forward. I fell off my seat, crash landing on the bottom of the boat. She laughed and dialed up the speed. We sped across the smooth surface of the lake like a skipping stone.

We finished our plots at about two o'clock. The sun was high in the air. The sky was cloudless. We stunk of sweat and Moskol.

The aluminum boat was so hot to the touch that it burned our skin, right through our coveralls. We took our lifejackets off and sat on them. It was the only way to make it bearable.

The faintest whiff of a breeze ruffled the lake. A wispy heat haze hung over the water.

Sandy yanked on the cord. The motor sputtered. She yanked again and again. The Evinrude refused to start. After the fourth try, it didn't even sputter.

I took over and tried some more, but to no avail.

"It's no use," I said, tiredly. 'It's flooded."

"What now?"

"Row," I said.

We stripped down to our string bikinis, picked up the oars and started rowing. After twenty minutes, we stopped and dipped our heads over the sides to cool ourselves down.

"Want to try again?" Sandy asked from the bow of the boat.

"Sure."

We tried several more times, yanking on the cord until our shoulder muscles rebelled, but to no avail. We started rowing again. We could see our campsite and our picnic table in the distance. That was all we needed for motivation.

I pulled hard on my oar, lifting my body off of the seat. The oar post snapped in two. I landed with a curse on my back.

"What else can go wrong today?" I whined as Sandy helped me up.

We clinked our last two cans of RC Cola together and guzzled them down. A couple hundred yards off to our starboard was a small rocky island. We had passed it earlier and even discussed moving our campsite there.

The island was composed of fault laden, grey igneous rock. Several stunted and gnarly spruce trees formed a natural wind break on the northwestern side. The top of the island was fairly level with bits of hardy sedge grass growing in places. It looked like a great place to pitch a tent so we decided to check it out.

We dove into the water, grateful for the coolness it offered as we towed the boat to shore by its mooring line. We hopped up onto the rocks and then placed a boulder on top of the mooring line to keep the boat from drifting away.

We lathered on the sunscreen and stretched out on the rocks, happy to rest for awhile. A stiff breeze kept the mosquitoes at bay. Before long we fell fast asleep.

I awoke to something crawling on my stomach. I brushed it away only to have it return with a flurry of scrabbling little legs. I heard Sandy curse and smack her face.

We both sat up, disgruntled.

The sun was beginning to set. The lake was a colorful rainbow of pinks and reds, the sun a ball of fire in the west.

"Don't look down!"

"What?" I answered, looking down.

It was my worst nightmare, worse even than drowning in the beaver pond, one that had plagued me since I was little.

The rocks were crawling with thousands of the largest spiders I have ever seen. They ran at us en masse.

We screamed and ran for the boat.

The spiders charged after us.

They were huge! Brown and furry with long legs and beady eyes.

We grabbed the mooring line from under the rock and vaulted into the boat. The boat skipped violently backwards under our momentum.

To our horror, the spiders didn't stop at the water's edge. They continued on, skating over the water towards us, some of them diving underneath it.

Sandy scrambled to the back of the boat and yanked on the motor's cord in a panic. The engine coughed, once and then twice, before roaring into life.

We sped away, the motor at full throttle, leaving a wave of spiders skittering across the water in our wake.

I haven't seen a spider to match these ones in size to this day and my fear of spiders has grown perceptibly.

🦋 🦋 🦋

We dragged the boat ashore, exhausted beyond belief. I knew we were both going to have nightmares that night. Who wouldn't?

We lit the Coleman lantern and proceeded to make dinner.

Sandy wanted Kraft Dinner and a bacon and tomato sandwich. I wanted beef stew. With only two burners on the Coleman stove, we ate in shifts and cleaned out our one pot so that the other person could use it afterwards. Rock, paper, scissors decided who ate first.

Paper covered rock.

I took my bowl of Campbell's beef stew into the tent and stretched out on my air mattress. I listened to the snap and crackle of bacon crisping in the fry pan. I hoped that the smell wouldn't attract any unwelcome critters…like bears.

I had a healthy respect for bears and kept my Peter Pan shoes in good working order.

I put my empty bowl aside and took up my needlepoint. I enjoyed reading, but sometimes wasn't up to concentrating on a story line, especially after an overly adventurous or arduous day. Needlepoint kept my fingers limber and settled my mind. I had already gotten a good jump on Christmas. I was almost done my mother's present, a complicated waterfall pattern with lacy fern fronds and knotted pink flowers.

Sandy finished her dinner and washed up before coming into the tent.

We chatted amiably about our families.

"You have a boyfriend?" I asked her.

"No one special," she answered.

"Really, I thought maybe you had a bit of a hard on for Jimmy."

"Get real!"

"What? Jimmy's okay. He's become much more likable since he has been working with Debbie."

"She's got him wrapped around her little pinkie now."

"Ah, so you do like him."

She gave me a dirty look.

"When are you getting married?" Sandy asked, changing the subject.

I chuckled, knowing deep down that I'd hit the mark.

"September 5th. I'm going back to college to complete a two year course in Geological Engineering. We're going to do two years in one."

"You really are whacko, aren't you?" she said, incredulous.

"Certifiable."

"Who am I going to work with after Labour Day then?"

"Maybe Ocean?"

"That'd be alright," Sandy added, staring off into the distance.

The drone of mosquitoes filled the night with the sound.

We heard the radio crackle. It was on the table outside.

"We forgot to check in."

"Blind River Base to Blind River Six. Over," echoed Ben's weary voice.

We were in trouble. Ben was on the line, not Ocean. We were hours past check in.

"You answer it," I pleaded.

"Coffee in bed for the next two days?"

"Alright,' I agreed reluctantly.

"And I get to cook dinner first?"

"No!"

"Wrong answer," Sandy finished. "You're the boss."

I struggled out of the tent and answered Ben's third call. Mosquitoes dive bombed me.

"Base this is Blind River Six checking in. Sorry, we had engine trouble today. Over," I said, not lying, but not telling the whole truth either.

"You still at Flack Lake? Over," Ben responded, sounding a little less annoyed.

"Yep. Over."

"John and I will be up that way tomorrow. Might be the plugs. I'll bring some new ones along. Over."

"Thanks, Ben. Can you bring some screws and a new oar mount too? One broke when we were rowing back to camp. What time do you figure? Over."

"Okay. We'll be there about ten a.m. Nightie-night, girls. Over," Ben closed off.

I ran back to the tent.

"Wonder what he and the big boss are doing coming up here," I said to Sandy, zipping up the tent's fly behind me.

"Maybe they're checking up on everyone and doing some lines."

"They do that?" I asked, shocked. I didn't think anyone checked up on us, but I guess it made sense.

"I heard one of the mechanics talking when you were signing for the truck and boat," Sandy said, hands behind her head. "I guess one of the crews have been caught fudging some statistics."

"No way! Which one?"

"Dunno. Maybe we'll find out tomorrow."

"Well, it's not us at any rate."

I settled back to my needlepoint.

"Bet you a dollar it's Sean."

"Why do you say that?"

"I worked with him last year, remember?"

I waited for Sandy to elaborate but she didn't.

I gnawed on my cheek. I wondered if we'd make a mistake on our lines. How lenient were they if we had? I wasn't sure.

Thoughts of getting fired replaced thoughts of charging hordes of carnivorous spiders for the night.

🦢 🦢 🦢

We slept in until eight and then got up to make breakfast. There was no point in rushing as we had to wait for Ben and John.

Sandy cooked up the last of the bacon and fresh eggs. I made myself a bacon and egg sandwich and crawled back into the tent to eat it. The tent was warm, but not stifling. Sandy sat by herself at the picnic table.

"OH-MY-GOD," Sandy said the words like a prayer.

"What?" I asked from inside the tent. I had finished my breakfast sandwich and was tidying up my back pack.

"Laura?"

"It's not a bear, is it?"

"Worse!"

I couldn't imagine what was worse than a bear?

"It's a skunk," she said, her voice quavering.

I opened the flap wider and saw the strangest of tableaus: Sandy sat at the table in mid-bite, a forkful of scrambled eggs was inches from her mouth, and a few paces away, a large skunk checked her out. He waddled towards Sandy. He stopped a few feet away from her and lifted his soft black nose, scenting the air. His nostrils and tiny whiskers quivered.

"What do I do?" she whispered frantically.

"Don't move," was all I could think to say.

The skunk padded towards her, tail down but the unmistakable sour odour of skunk revolving about him in a cloud of bad cologne.

"Laura, he's crawling up my leg," Sandy croaked in terror.

The skunk scratched at Sandy's leg like a beagle looking for a treat. The little fellow had obviously been visiting the campground for some time and had been getting handouts. After all, who could refuse that face and those two white stripes down that silky smooth black back?

I choked down a laugh. It really wasn't funny. Okay, it was, even if we did sleep in the same tent.

"Where's your camera?"

"Out here," she said through clenched teeth.

"Oh, that's too bad," I said, unable to keep a straight face. "You better drop him something."

The skunk walked around Sandy's leg, examining her. He climbed up first one and then the other leg, scratching lightly.

Sandy turned her fork upside down and the scrambled egg fell to the ground. The skunk nibbled on the scramble eggs. He looked up, wanting more. She dropped a piece of bacon at her feet. The skunk waddled off with the bacon hanging out of either side of his mouth.

"That was close."

"You are soooo lucky," I agreed.

I zipped the tent up and went back to my sorting and rearranging. As I was putting my clean clothes away, I heard a scratching at the side of the tent.

A short shadow skirted around the nylon tent, little claws pattering at the tent flap. I knew instantly who it was.

"Sandy! He's back!"

"Hah, now you don't think it's so funny, eh?"

"Your gear is in here too."

Too late, I realized I hadn't done the fly up all the way. I dived into my sleeping bag and pulled the covers over my head.

I lay there, trapped. Little claws pitter-patted against nylon. It seemed like an eternity before I was brave enough to look out from under the covers. There was nothing there.

I stood up and peeked outside.

Sandy lifted her coffee cup in a toast to me.

"Is he gone yet?"

"Ages ago."

"Why didn't you tell me?" I groaned, irritated.

"Payback's a bitch."

We burst out laughing and decided to name the skunk Maurice.

A stern faced Ben and John drove up about an hour later. They were accompanied by a third man who we didn't know. It was clear to us that they were in a very bad mood so we regaled them with the story of Spider Island and our breakfast guest.

"Those were dock spiders. There are lots of them in northwestern Ontario," John said. "They are the largest of the fishing spiders, called *Dolomedes*. They can walk on water and dive underneath it to feed on small aquatic insects and fish."

"You guys are lucky. They have a really mean bite," Ben added.

The other guy didn't say much.

To me, he looked out of place. His boots were new and his khaki pants were freshly pressed, as was his olive green forestry shirt. I noticed two red hard hats and one orange one in the pickup truck that they drove in with. The orange one was Ben's, meaning that this guy was a big cheese of some sort.

"So you girls are having engine problems," was the first thing the stranger said.

I thought it funny that neither Ben or John introduced him.

"Yeah, I think it's seizing," I countered.

"Well let's let the men have a go at it, shall we?" he said.

I hit the roof! 'Let the men have a go...'

"I don't know who the fuck you are but I worked Creel Census last year and have a full forestry diploma. I can strip down a motor as well as any of the mechanics at the shop. I've already cleaned the plugs, the fuel lines and reset the pull rope. The engine only works intermittently, if we are lucky. I've already broke one oar post rowing across this frigging lake, which if you haven't noticed, isn't that small."

Sandy stood back and silently watched me vent. I was mad. Who was this idiot?

John looked like he was going to start tap dancing and Ben was making slashing motions across his throat.

"Well, I didn't mean to hurt your feelings, just that you should let a man have a go at it," was how he finished.

"I don't need a man to fix my engine!" I countered.

John walked over to our boat and lifted the motor out of the water. He yanked on the cord a dozen times. The motor refused to turn over.

Sandy looked at me and shrugged.

"Like I said, it's seized."

"It's probably been dropped in the brink," John volunteered.

"Wouldn't whoever had it last report that, John?"

"Not always," John returned.

He grinned at me in passing.

Ben pulled a another 15 hp Evinrude out of the back of their truck while John unhooked the motor that wasn't working.

"What about the oar post?"

"Behind the seat in the truck," Ben answered.

Sandy went and retrieved the oar post.

"Thanks for the new motor," I said.

"Guess we better get at it," Sandy added.

"We've got lots of work to do still," I said, glaring at the stranger.

"I'm told you gals are doing a great job, better than some of the boys so we'll let you get back at it, as you say," the stranger finished.

"Alright. You girls be careful out there," Big John added.

We watched the men drive off.

"They fired Sean, I knew it!" Sandy exclaimed.

"We don't know that."

"Yeah, we do. Sean got caught cheating last year. He marked a stand of poplar as a stand of spruce. Big difference. He got a warning and was lucky to be hired back this year."

"Wow!"

We shook our heads in disbelief and packed the boat up for the day.

"Wonder who that doofus was?"

"Whoever he was, he's going to remember you for a long time," Sandy said with a chuckle.

"Guess I got carried away. I just hate condescending men."

"No kidding."

It was a sobering afternoon. I found out later that the stranger was a Senior Director of the Ministry of Natural Resources from the head office in Toronto. Whoops!

🌱 🌱 🌱

On our last day at Flack Lake, we fried up some spam and made a big pot of scrambled eggs from dried egg powder. We made three plates, one for Maurice and one for each of us.

Maurice came for breakfast every morning and sometimes stopped by for dinner. He seemed to even recognize the sound of our truck coming home at night when we drove back from our plot lines that were farther afield.

We were going to miss him; Sandy the most. The pair of them had developed quite an unlikely friendship.

We said our goodbyes and packed up the tent. Maurice hung around, as if he knew we were leaving. We left him an apple to nibble on.

As we drove out of the campsite, we were met by two campers pulling boats and trailers behind them. They waved to us. We waved back.

"Don't forget to feed Maurice. He likes his bacon extra crispy," we shouted out the window.

We were met with perplexed looks.

That was okay as we knew Maurice would be finished his apple by the time they arrived.

Chapter Six

Lost In Translation

"This is it," Ben said, pulling up beside an impregnable old logging bridge that looked like it had been in use since before Davey Crockett was born. The sleepy little sun-dappled river called White River that flowed beneath the bridge was silt bedded and inviting.

We hopped out of Ben's truck and unloaded the sturdy red canoe that we had used previously on our fly-in. We dropped the canoe into the water, tied the mooring line to a rock, and loaded it to the brim with all our gear.

"Don't get dumped this trip, okay?"

"Hey, we didn't the last time," Sandy admonished him.

"You two just seem to find trouble wherever you go."

"We can't be the only ones who have been chased by spiders, got up close and personal with the grand-daddy of all black bears, and adopted by a skunk," I said innocently.

"And your dad was really nice too," Sandy added.

"That's right, he was," I agreed.

"Just stick to the river, ladies," Ben said with a helpless shrug, "and I'll pick you up in six days at the bridge off highway 106."

"We'll be there."

We stepped lightly into the canoe and Ben pushed us off with a wave goodbye.

"No more adventures, right?"

We waved our paddles in the air in response.

We deftly maneuvered our canoe down the little river. Sandy was paddling like an expert now. I still did all the steering. We headed downstream so the current carried us along at a good clip.

Sandy paused and put away her paddle. She pulled the aerial photo of the river out of her pack and examined it closely. I let the current take us, correcting our course only as needed.

"There should be an open patch of ground to our left around the bend ahead. We can pick up two plots from there," Sandy said, tucking the photo away.

"Sounds good to me."

We banked the canoe on the small gravel shoal we found around the bend. Sandy tied the canoe's line around a small alder, hardly four feet high, and then gently pushed the canoe into the current while I tied an orange marking ribbon around another alder. Once again, our food was safer out on the water than on the land.

Sandy took a compass bearing and we walked off into a small clearing teaming with bushy horse tails and hardy swamp grasses. The spongy wet ground almost sucked the boots right off of our feet. Gobs of mud and clay clung to our Grebbs by the time we reached the other side.

We climbed a long slanting hill. The plot we were to traverse started at the top of the hill. We found ourselves in a stand of ancient spruce trees. Their bark was thick and scabrous. It was an easy plot to complete and we were finished in less than an hour. The next plot wasn't so easy.

A stand of spruce, balsam and birch greeted us next. The ground was choppy, filled with rotting logs and stumps. We may have been fresh from four restful days off, but the going was treacherous and we were soon tired and irritable.

Flies hummed about our heads. Our hardhats glistened with bug repellant, but that only kept the mosquitoes at bay. It didn't deter the biting hornets and deerflies.

I took the last core sample that we would need to complete this line, and then recorded my findings on my data entry sheets. Sandy looked for signs of insect infestations and disease and recorded those.

"I'm done, want to pitter patter, Peter Pan?"

"Did Peter pick a pack of pickled peppers?"

"No," I said, "he put on his Peter Pan shoes and flew!"

Sandy laughed as she took a compass reading. "Let's fly, Lol," she said.

I strode after Sandy, but was having a hard time keeping up with her. While she was shorter in stature than I was, she had legs that went on forever and was much faster.

We climbed through a small hollow and over a decayed ant infested log, pushing away tearing vines as we did so. Over to our left was a rocky incline. The sound of water tinkling into a pool caught our attention. A waterfall tricked down from the steep walls of the cliff and into a small sandy pool at its base.

It was stifling hot.

We were both girls!

We looked at each other and grinned.

We dashed to the waterfall, stripping as we went. The water was cold and refreshing against our bare skin. Tall lacey ferns ringed the pool. Only a few birds called in the forest. A whisper of a breeze rustled the leaves overhead. It was heaven.

We lay there for some time, letting our bodies cool and minds drift. After a while, we realized we had to get going, we still had to find a place to set up camp down river somewhere.

Reluctantly we climbed out of the pool and pulled on our sticky shirts and jeans. We had left our coveralls in the canoe as it was still early in the day and we hadn't thought it was going to take this long to do two plots.

Sandy took a sighting on the compass and we headed off, back towards our waiting canoe.

We struggled over slash and logs. Sandy took a wrong step. Her right foot sank deep into the rotted pulp of an old spruce tree that had fallen to fire decades before. A huge swarm of black and yellow hornets blanketed the air between us.

"RUN!" I screamed.

Sandy yelled out in pain and beat furiously at the back of her shirt. I grabbed her by the arm and yanked her forward.

"Run," I screamed again. "Hornets!"

We bolted in two different directions, hoping the swarm might be confused. It wasn't! The swarm split up and followed us both.

I saw Sandy tear her shirt off as she ran. She flailed at the angry procession with it and yelped in pain as they responded in kind.

We both ran breathlessly through the bush, unheeding of the brambles and thorns that tore at us. We were of singular mind. Get to the water!

I stumbled and almost fell. A couple more hornets stung me. It hurt like the dickens.

We ran blindly on, hoping we were heading in the right direction. All at once, we opened on the bit of swampy ground that we had started from and made a bee-line for the river.

We dove into the water, a dozen or so yards away from each other.

From underwater, I looked up saw the clouds of hornets furiously circling over my head. I held my breath as long as I could. When my lungs couldn't stand it any longer, I took a quick breath of air and submerged myself once again.

It seemed to take forever before the hornets finally flew off.

I met up with Sandy. She was pale and shaky.

"You okay?"

"I feel sick."

"You aren't allergic to them, are you?"

"No, but I got stung lots," she said.

Sandy showed me her back and neck. She had at least six swollen bites.

"Hold on, I've got to get the first aid kit out of the canoe and get some Sting Stop on that," I said.

"Yeah, okay," she said faintly.

I dashed for the canoe, glad to see it was where we left it. I pulled out the med kit and grabbed some antibiotic appointment and Sting Stop.

I treated Sandy's swollen bites and helped her to the canoe.

"I think we better carry some antihistamine and Sting Stop with us when we cruise," I said as we headed down the river.

"Good idea," she whispered, pale and drawn.

Sandy took an antihistamine and downed a full bottle of water.

We found another sand bar a few hundred yards down the river and decided to pitch the tent for the night. We'd make up the time the next day. Sandy needed to rest to recover from all the hornet venom.

So much for staying out of trouble.

🦢🦢🦢

The next day we lay sunbathing on a flat, long ridge of basalt rock overlooking White River. The forest below and behind us was quiet. Not a bird chirped or breath of air disturbed the peacefulness of the afternoon.

The river meandered below us like a piece of silk that had fallen off of a loom. The green and gold canopy of the forest stretched away from its borders as far as the eye could see creating a rich tableau of light and dark. Smaller ridges and hills rose and dipped amidst the tapestry. Heat waves shimmered, creating a silver haze. Even the mosquitoes and horseflies didn't dare disturb nature's seamstress.

I let the heat exhaustion drain from my body and into the rock at my back. I felt like a wilted piece of lettuce. We had risen at four thirty in the morning and started work by five thirty am.

Sandy took some photos of the view and then we both dozed for an hour until Sandy woke me with a shake.

"Hey, Lol," she said, the shake turning to a poke in the ribs.

"What?" I asked, shading me eyes from the sun.

"Look down there," she pointed.

I rolled over and looked to where she was pointing. There was a little red dot floating slowly down the river.

"You did tie up the canoe, right?"

"No, you said you'd get it!"

"I didn't say that."

"Yeah, you did when I was ribboning off our starting point."

We both groaned and stood up. Every joint ached with fatigue. We donned our packs and coveralls and hurried down the hill.

Everything we owned was in that canoe. We packed up our campsite in the morning and headed down river. We then stopped briefly for a quick lunch of canned ham sandwiches and cold RC Cola after two plot lines. We hadn't found a new campsite yet, deciding to continue working instead.

Silently, I knew we were each blaming each other for our stupidity.

We settled into a fast shambling trot down the hill, not wanting to step in any more hornet nests. We reached the river, but weren't sure if we were ahead or behind the canoe.

We discussed splitting up but decided against it. You have to have each other's backs. That doesn't work when you can't see the other person.

As we stood stupidly wondering what to do, the canoe floated by us. It spun in lazy circles on the current. We ran downstream after it, but it kept getting ahead of us until finally it got caught up in an uprooted tree stump.

"Who's going to swim for it?"

We eyed the river warily. The river was deep and cold. It was also running fast.

"Alright, I'll go," Sandy finally volunteered.

"We should wrap a rope around your waist and I'll hold it in case the current is too strong," I said, before realizing that all our rope was in the canoe. "Okay, right, that won't work."

Sandy raised an eyebrow at me as she got down to her skivvies and waded into the water with her steel toed boots still on her feet. The mud sucked at her like a baby looking for a tit.

Sandy reached the canoe in a few swift strokes, the river carrying her towards it. She labored to swim back, tugging the canoe behind her against the current. She emerged from the river dripping mud. A giant leech was attached to her back. I pulled out my Bic lighter and burned it off her. It died with a hiss and a convulsion.

We paddled half-heartedly down the river looking for a suitable place to camp. We spotted a small crescent shaped gravel beach with some sedge grass at the edges. There was just enough room for the tent and a tiny fire pit. It would have to do.

We pulled ashore and leveled off a section of beach before pitching the tent. We then scooped the gravel into a ring and made a nice campfire to roast some wieners and marshmallows on.

It ended up a pleasant evening. The smoky tendrils from the burning limbs and dried needles in the fire pit, along with a tiny piece of green coiled Pic that we had lit, kept the bugs at bay. The campfire crackled and popped. Our shadows danced across the tent. The river gurgled. Stars by the quad-drillion shone down upon us. It was primitively serene in a soul-satisfying kind of way.

The wind brushed the glowing coals into ever-smaller dwindling flames as the night lingered on.

Sandy yawned and wished me goodnight.

I sat for awhile staring at the stars wondering who was staring back at me. I thought of God, the courage of the boy with one leg running across Canada, at first unheralded but now with a cheering entourage. I wondered too if I wasn't too young to get married. I wondered also about the person who had painted the 'No Whites Allowed' sign on the train trestle and what had happened to them.

With one last look into the fire's dying embers, I grabbed my hardhat and squatted down by the river, filling the hat up with river water until it over-flowed. I then poured the water over the dying embers and stirred it into a gooey wet mess with a stick. I didn't want forest fire starter on my resume.

I crawled into my sleeping bag. Sandy was fast asleep, purring lightly in her dreams. Sleep came to me fast too.

Around two in the morning, Sandy and I were startled awake by a frantic call for help over the hand radio.

"Mayday, mayday. This is Blind River Crew Five calling anyone on the Forestry channel. Please respond. Over." It was Crew Five, Jimmy and Deb.

Sandy picked up the mike to respond when the radio immediately buzzed into life.

"Blind River Crew Five, this is Base," a sleepy Ocean said. "What's the emergency? Over."

"We're being attacked by a bear," came the frenzied response.

'What?" Ocean asked, no longer sleepy.

The following is a recount of what happened as told to me by Debbie:

A few yards from camp, a graying black bear rummaged around a fire blackened stump in the forest. He tore it open with his sharp claws, looking for grubs. Grunting, he shredded the pulpy wood and gobbled down the ants and larvae he found there. At six hundred and fifty pounds, he needed a lot more than ants and larvae to satisfy his hunger.

His paws were pidgeon-toed, the knuckles crooked with arthritis. He approached the campsite warily. He stood on his hind feet and sniffed the air. He recognized the smell. It was the smell of people and people had food.

At first, he rifled through the loose earth by the cold fire, digging out the steak bones that had been buried there. He broke those open with one chomp and ate the marrow.

He then found a couple of empty tins of beans. He stuck his tongue inside the cans and licked out what little remained of the sweet tomato sauce. He grunted loudly when he was done and crushed the cans underfoot.

Debbie awoke from a strange dream about clanging cymbals and grunting clowns. She listened to the night and realized that she hadn't imagined hearing all that grunting. There was an awful lot of it coming from outside the tent. She reached over and woke up her partner.

"Come on, Jimmy, wake up," she whispered urgently.

Jimmy awoke instantly.

"What?" He asked, sitting up.

"Listen," she said.

There was no moon that night. It was pitch dark outside the tent. Crickets chirped. Frogs croaked. The bear snuffled and grunted. And then the night went silent.

"Shit, it's a bear," Jimmy said.

"No kidding," Debbie retorted, pulling away.

They unzipped their sleeping bags as quietly as they could. Jimmy grabbed the hunting knife that he always kept under his makeshift pillow. The sheath caught on a strap and he struggled to get it out. Debbie snatched up the radio, clinging to it like it was a life preserver.

The wind rose. The tent snapped and fluttered on its tether. The bear growled. The tent continued to snap and flap. The bear growled louder. A sudden gust of wind picked up the tent and tossed it sideways like a cowboy on a bronco's back.

The bear attacked

Giant claws ripped open the tent's thin nylon fly cover.

"Go, go, go," Jimmy yelled, slitting open the nylon wall behind him.

Debbie and Jimmy dove threw the makeshift doorway.

The bear growled and savaged the tent behind them.

They vaulted into the canoe, bare feet sliding on the slick wet metal. Jimmy cut the tow line and the canoe rocketed out into the river as the bear sliced the tent into quarters and then eighths.

The pair of them huddled half-naked in the canoe listening to the bear's rage as he shredded everything he could find. Jimmy's bare chest shone with sweat. Debbie clutched his hand in a death grip. Terror filled their hearts. Jimmy wrapped an arm protectively around her shoulder.

They watched helplessly as the bear tore through their belongings, ripped open their cooler and emptied out their dry box of supplies. He pried open the metal cooler like it was aluminum foil. Once finished with the milk, eggs and bread, he discovered the boxes of Raisin Bran and Kraft Dinner.

After an hour, he got tired from all of the excitement and the big dinner and wandered off into the forest.

Debbie and Jimmy stayed huddled together in the canoe for some time.

"You got the radio, right?"

"Yeah," Debbie said, her teeth chattering.

Deb handed Jimmy the radio and he unclipped the microphone. They had a hard time checking in earlier because they were in a river valley and the reception was poor.

"You're going to have to stand up and hold up the antenna," Jimmy said.

"Okay, but do you think anyone will be listening?" Debbie asked, one hand on Jimmy's shoulder for balance. With the other, she held up the antenna.

The canoe rocked gently under her standing weight.

"Hope so," he said. "Mayday, mayday. This is Blind River Crew Five calling anyone on the Forestry channel. Please respond. Over."

"Blind River Crew Five, this is Base," a sleepy Ocean said. "What's the emergency? Over."

"We're being attacked by a bear," Jimmy whispered frantically into the mike.

'What?" Ocean asked, no longer sleepy. "Hang on and I'll ring Ben up. Are you okay? Over."

"Yeah," Jimmy said, "we're in the canoe. The bear shredded our tent and all our gear. Over."

The radio crackled something unintelligible and then went dead. Jimmy checked the other channels, but there was no response. Debbie sat down with a thump.

"Now what?"

"We keep trying," Jimmy answered.

After thirty minutes of having their call for help go unanswered, they gave up trying. The radio's battery was so close to dead that trying made no more sense.

They huddled together for warmth. There was nothing more to do except stay as warm as they could until the sun came up.

Debbie watched the sun come up with mixed feelings of joy and apprehension. The apprehension won out.

There was carnage everywhere.

She and Jimmy struggled to find something to wear. Debbie found a one legged pair of jeans which she slipped over her cotton briefs as well as a torn t-shirt. Jimmy found his shorts, but everything else was done for. They searched through the campsite and several paces into the forest until they were able to find their boots. It was the best they could do.

"What are we going to do now?" Debbie asked, deflated.

"We'll paddle down to our pick-up spot. Hopefully Ben will send someone out with a truck looking for us, maybe a chopper."

'Okay," she whispered, looking at the garbage strewn about the campsite. "What about all this?"

"We'll pack as much as we can into the canoe and bury the rest. That's all we can do."

The pair worked tirelessly, burying what they couldn't take out. They ate a couple of cans of cold soup for breakfast, and then headed downstream in the canoe.

They paddled down river as fast as they could. Every so often they thought they heard the sound of a helicopter overhead and waited, but nothing appeared.

"What happens if Ben isn't at the bridge?"

"We'll flag down the next truck we see and hitch a lift," Jimmy said earnestly.

Debbie started to laugh.

"What's so funny?"

"Look at us. You're shirtless with your boxer shorts showing through the ripped pocked of your shorts and I'm wearing a bear goobered pair of one-legged jeans and a shredded t-shirt."

"I see what you mean," he giggled.

They burst into laughter and continued paddling until they reached the bridge where they prayed that Ben or someone else from the office would be waiting for them.

The river they were on was quite narrow and windy. Willows and alders draped leafy arms over the water. Sometimes, Debbie felt like she was canoeing through the bayou.

They reached the bridge with little difficulty and pulled the canoe ashore and tied it securely. There was nothing of value in it, just broken equipment and shredded clothing.

Jimmy helped Debbie climb up the bank.

There was no one waiting for them. No Ben. No Big Jim.

A couple in a Winnebago saw the disheveled young couple standing at the side of the road and pulled over. It was clear the young couple needed help.

The Saunders family were wonderful, Debbie thought. They drove Jimmy and Debbie right up to the park office.

Within twenty-four hours, Smokey, as the bear was now known, had become a celebrity. Everyone was on the look-out for the old grey bear. He was sighted in fifteen different places at once. Jimmy and Debbie's tale got embellished into the tallest of improbable tales with both of them being left without limbs and clinging to life in the hospital. There were some other steamier tales about their naked night in the canoe, but we'll let you, the reader, decide about that on your own.

Ben picked them up and drove them back into Blind River, giving each of them a hundred dollars to buy some new clothes. Within forty-eight hours, they were back in the bush on a fly-in.

The rest of our four days on White River were uneventful. Our friends were safe. We slept within a few feet of our canoe every night. We were ultra-happy that we didn't cross the old grey bear's path as we were in the same general area that Deb and Jimmy had been in.

Ben assigned us to the park after our four days off, begrudgingly handing us the keys to his truck and the Terrajet. We were ecstatic!

Chapter Seven

Mad Dogs and Englishman

Summer came in like a lion.

The forest was tinder dry.

Between twelve and three p.m., one could have heard a pin drop for miles in either direction, the silence so complete under the relentless heat that birds, bees, beetles, and mammals of all sorts curled up to sleep beneath whatever piece of shade they could find. We wanted to be no different, but had a job to do.

The day was sticky and humid. Our movements were leaden and snail-slow. We prayed for rain, but knew it would be at least thirty days in coming.

The smell of sour sweat stank up the cab. Even with the truck windows open, I could smell it coming off of our bodies. The wind whistling through the window provided no relief from the scorching day.

Sandy drove, alternating the gears into four-wheel drive expertly as the old Dodge Power Wagon kicked and bucked its way over the steeply rutted bush track. Visions of a cold mug of Blue flashed before my eyes. I licked my lips, my mouth watering. Instead of the sweet taste of beer, I found only the bitter taste of salt and parched skin.

I leaned limply against the seat, my bare arms sticking to the vinyl bench seat, my eyes half-closed. I dreamt I was in my parents pool floating lazy circles on an air mattress, my mom swimming laps around me. She stopped and leaned against the the air mattress, sending a cool burst of water over the sides. She laughed and then began to sing the chorus to one of her favorite English songs. I quickly joined in.

Mad dogs and Englishmen go out in the midday sun.
The Japanese don't care to, the Chinese wouldn't dare to,
Hindus and Argentines sleep firmly from twelve to one,
But Englishmen detest a siesta,
In the Philippines there are lovely screens,
    to protect you from the glare,
In the Malay states there are hats like plates,
    which the Britishers won't wear,
At twelve noon the natives swoon, and
    no further work is done -
But Mad Dogs and Englishmen go out in the midday sun.

"What on earth is that you're singing?"
"What?"
"That song? What is it?"
"Oh, sorry," I said, "I didn't realize I was dreaming out loud. It's a song by Noel Coward. My mother loves to sing that song when it gets so damn hot like this."
"And you know all the words."
Did I catch a shot of diffidence in her voice? I thought so.
"Just the chorus. My mum produces and directs music hall theatre. It's one of the songs they do in her show sometimes."
"Have any others?"
"I don't. She does. She knows songs about Lizzy Borden and her axe, gay caballeros, and poisoning pigeons in the park." I smiled. They were some of my favorites too.

"My mum even had a song written about her. During World War II she used to call the troop trains in at Woking Station, and then after the war, a fellow wrote a song about the Golden Voice at Woking Station. If you're ever in Ottawa, I'll get you in to see one of her shows. They're really funny. *Plus Six in Song* is her group's name."

Sandy was never going to come to Ottawa to visit me and would never see one of my mum's shows. We both knew that. Friendships in the Ministry rarely lasted for more than a summer. It was just how it was.

We bumped along in silence. I couldn't get that song out of my head. It became quite annoying after an hour or so. I closed my eyes and tried to think of another.

The old Dodge dug its way up a steep hill. With a thump and a sharp crack, we hit a huge pothole. My head slammed into the ceiling.

"Shit! That hurt!"

"Poor baby," Sandy said sarcastically, leaving me to wonder if she had done that on purpose.

I reached for my yellow hardhat and popped it on my head. The next pothole didn't hurt so badly.

Sandy swore as we topped the hill, the truck nearly stalling. She hit the gas and it lurched forward in a hail of gravel and spinning tires.

I looked out the dusty window: a sun-dappled golden lake nestled between a row of hills within a cigar shaped valley. It looked surreal. The heat haze made the lake and valley appear like a mirage, the lines between lake, shoreline and sky a blur of colour. This would be our home for the next ten days.

We arrived at the end of a weed covered dirt track to where a wide sandy point dipped off into the cool stillness of the lake. There wasn't a breath of wind to disturb the water. No flies dive-bombed the water's surface or hummed amidst the sand or the sedge grass that bordered it. No lonely loon calls broke the quietude.

We meticulously set up our camp, pausing now and then to wipe the sweat from our eyes and look out at the haunting beauty that surrounded us. The air was oppressive. Every sound we made was dull and hollow like a hand slap on the inside of a cardboard box.

We didn't speak much; it was too much effort. We silently knew that we weren't going to do any plots today. We'd taken to rising with the grey dawn at 4 am and began work by 5 am in order to beat the heat. Tomorrow would come soon enough.

After we finished setting up the tent and blowing up our air mattresses, the lake's call was too loud to ignore.

We stripped naked and dove into the water. The water felt like cool silk against my skin. I swam out about two hundred yards from shore and reveled in the cold caress of the lake.

Sandy launched out from shore on an air mattress. She donned her sunglasses and pulled her baseball cap over her eyes and floated beneath the hot summer sun.

I smiled. There were advantages to working in the bush with another woman, miles from civilization.

I swam back to shore, caked myself with sunscreen and dove onto my own air mattress. It sped across the water like a skipping stone.

We spent the next several hours lazing the day away.

An uneasy silence had developed between us over the last couple of weeks. Work had become humdrum and we had run out of things to talk about.

Sandy didn't want to hear anymore about my fiancé and I didn't want to hear anymore about her lack thereof. She had gotten over her crush on Jimmy, which was a good thing, as I had the feeling that he and Deb were now an item, albeit a quiet one as family and spouses can't work together in the Ministry. Sandy had become rather morose and sarcastic. The only real conversation that we had was in the truck and it had taken a crazy Vaudeville song to start it.

"I hear you two aren't getting along so well," Ben had said to me in private before we left.

"Where did you hear that?"

"From Ocean and Debbie," he said. "I need you two to work it out. We're down a crew since we fired Sean and with what happened to Ocean and Colin, it doubles the effect."

"Don't sweat it, Ben," I consoled him, "we've only got a couple more months and it isn't that bad. I've got no tan lines, want to see?"

"No, I don't want to see. I don't fancy having that fiancé of yours coming at me with a twenty gauge."

We both laughed, but I hadn't been amused. I hadn't said a word to either Ocean or Debbie about being unhappy with working with Sandy so that left only one option.

It started to get cooler around seven pm and we made a nice dinner of bbq steak and mashed potatoes with mounds of butter on them. The coolness of the evening lifted our spirits.

We went to bed with the sunset. The nightmare that gripped me was paralyzing.

The sides of the tent fluttered softly in the afternoon breeze, the sun casting dark and light green checker board patterns on the thin nylon walls. Every sound was amplified. I heard the spider spinning its web in the corner of the tent above my head, the drone of a thousand mosquitoes bumping against the tent screen, Sandy's rhythmic breathing as she slept, and the rapid guttural breathing of something monolithic outside the tent.

What was that?

I bolted upright.

A black pall fell over me. For a fleeting moment, the world was gone, non-existent. A great empty darkness prevailed. I shivered uncontrollably. After a moment, the shadow was gone and the sun once again shone down upon the tent.

I sidled across the tent and peeked out of the tent fly. The truck was still there. The water was calm; the beach undisturbed. Our stove and cooking equipment was perched on the same big log we had used to cook on. Everything was as it should be.

I walked outside. The sunlight was blinding. I shaded my eyes. Behind me came a loud grunt. I turned slowly.

The black bear stood on his hind legs, drool dripping from his exposed teeth, snot running from his nose. His tongue lolled out to one side. His right side molars were broken off, the ragged edges looking raw and painful. His eyes were dark and feverish.

Sandy stood behind him. She smiled at me as if she knew I was about to die and that was okay with her.

I took a step backwards.

Could I make the truck? Was it even unlocked? I thought so.

In slow motion, like a stop-frame piece of film unwinding, I ran towards the truck, but it seemed to get farther and farther away from me the faster that I ran. I reached for the door handle, but it slipped from my grasp.

"Sandy, help me," I called to her.

There was no answer.

Hot searing pain tore through my left leg. I swung at my attacker. The bear let go of my thigh, his teeth dripping with my blood. I pummeled him in the face with one fist while trying to open the truck's door with the other hand, but the door was locked.

The bear slashed me across the face. A waterfall of blood covered my eyes.

I screamed.

I looked over the bear's shoulder and saw Sandy wave good-bye to me through a bloody haze. She smiled and then went back inside the tent.

Sandy and I sat bolt upright at the same time, screams echoing off the tent walls.

Sandy flicked on the flashlight.

"That was a bad one," she said, gasping for breath.

"Yeah, me too," I agreed, letting out a sigh of relief. It really had been a doozie.

"Must be the heat," she said.

"Yeah, it's stifling in here," I agreed.

We lay there in silence, neither of us wanting to talk about it. I idly wondered what her nightmare had been.

Before I knew it, our wind up clock's alarm was piercing the morning stillness. The alarm ran on and on until all life drained out of it when neither of us reached to turn it off.

Sandy finally got up and went to make coffee.

I laid in the tent and stared up at the ceiling. There were no spiders spinning webs above my head like in the dream, no hulking shadows other than Sandy's flickering across the tent walls.

I wondered if the dream had been sparked by my conversation with Ben? My trust in Sandy was shattered. I didn't know what I had done to make her want another partner. Should I ask her or just bite my tongue? I didn't know what to do.

The smell of sizzling bacon was intoxicating. The aroma had a life of its own.

I slipped my coveralls on over my bathing suit and went down to the lake to wash my face and teeth.

"Coffee's ready and bacon's done," Sandy said, breaking a couple of eggs into the fry pan. "How do you want your eggs this morning?"

"Over medium, I guess."

I sat down on the log that was our eating table and chair. This was a surprise. We hadn't made breakfast for each other for three weeks.

"You're in a good mood this morning," I said.

"It's going to be a great day," she replied in a Scottish drawl.

"Okay," I said skeptically, taking the plate of eggs and bacon from her.

We ate in silence. Afterwards Sandy washed the dishes while I made Spam sandwiches to take with us.

By the time we finished, the sun was rising over the hills and the morning mist had dissipated over the lake. We had one last cup of coffee and readied ourselves for the day.

"That was a wicked dream you had last night," Sandy commented as she packed away our lunches.

"Yeah, it was. Scared the life out of me. I dreamt I got tore apart by a black bear and you did nothing to help me."

"No shit!"

"No shit," I agreed.

"You aren't going to believe this, but I had the same dream."

"I ran for the truck, but it was locked," we said in unison.

"And the bear got me...," I added.

"As I was reaching for the handle...," Sandy finished.

"Weird."

We shook our heads in disbelief as we threw our packs into the back of the truck. Conveniently, neither of discussed the part where we disappeared into the tent without a backwards glance or offer of a helping hand.

"What do you think Freud would make of that?"

"Do we really want to know?" I countered.

"Nah," we said together.

We drove out of the campsite in contemplative silence. Maybe we both had trust issues, I reasoned.

We followed the track we came in on. A half mile down the road was another old logging trail that would lead us to the start of our first plot lines for the day.

We plowed over small saplings, using the Dodge Power Wagon's big grille to push them over. The skidder trail hadn't been used in years.

We looped around a bend and drove slowly past a dried up swamp. The muddy bottom was dry and cracked; the few pools of water that were left were shallow and stank of sulfur. A pair of beavers ambled along the right side of the road, flat tails dragging in the dust behind them, obviously looking for a new place to call home.

Sandy pulled up beside them.

I hung my head out the window and looked down. The beavers stopped beside us. The large male regarded me quizzically and then became fascinated by his reflection in the truck's metal rims. The smaller female soon joined him.

"Get your camera out, Sand. Come take a picture looking down at the beavers. They're looking at their reflection in the truck rim."

Sandy retrieved her camera from her pack and then leaned over me with the camera. The camera flashed as she snapped off a couple of shots.

"Oh, my god, they are so cute," she purred, snapping a few more pictures.

The beavers looked up at us and then the big male sunk his teeth into a tire.

"Drive, drive," I yelled, laughing.

Sandy pulled forward quickly, but was careful not to run them over.

"I think we made him mad," I said, not having ever left an angry beaver in my wake before.

We pulled over for a few minutes and laughed ourselves into a pee break. We checked the front tire to make sure it wasn't leaking and then drove on to our starting point.

Sandy took a bearing and we crashed off through the bush, startling a flock of grouse. They flew madly in every direction. Crows cawed in distress from the upper reaches of a poplar tree.

"Ain't no bear going to sneak up on us today," Sandy quipped.

"Not on your life."

"Or yours," she agreed.

"Too bad I don't have an extra McCulloch chainsaw t-shirt."

"Why's that?

"The slogan on it is 'I've got Beaver fever'."

We laughed outrageously and made noise, lots of it!

🐦 🐦 🐦

The following days were enchanting. Our mornings began with molten sunrises and our days ended with dragon-fire sunsets.

Whatever had brought on our tour's first night's joint nightmare had softened our relationship woes. We worked as a team again.

The day previously, we had taken the canoe out and completed a line at the end of the lake. We portaged a short ways to another lake and completed another line.

Approximately a hundred yards in from that smaller lake was a tall rock outcropping overlooking the water. We took our lunch break there.

Something white and out-of-place caught our attention farther up the rock incline.

"What's that?" Sandy asked.

"Dunno."

We wandered over, cans of RC Cola in hand, and took a look.

Stretched out on the rock, held down by two rocks was a sun-bleached white t-shirt. On the front was stenciled in black: **You touch'a the shirt, I break'a your face.**

We started to giggle. The giggles turned into an avalanche of laughter.

We picked up the shirt and tucked it away in a pack. This shirt deserved a place on the wall in the cruiser's office alongside the horse's skull and moose rack that the other teams had brought in.

The work got easier after that, despite the heat stroke that we battled daily.

The city slicker within me was now completely gone. The forest had become a part of me. I had traded middle class suburbia for nature's beauty. I didn't feel like a stranger in a strange land anymore.

I loved the dusty smell of the moss and leaves that littered the ground I walked on, the sweet scent of the wind over the lake, the gentle sigh it made in the leaves overhead, and the soft sound of birds chatting in the forest.

Ahead of me, Sandy tagged a white birch tree. She checked and rechecked the aerial photos and the base map to make sure we were in the right place.

We both took a bearing from the allocated plot and set our compasses. We marched into what should have been an uneventful traverse.

Sandy barreled through the bush and I barreled after her. We counted off our paces and I spot checked Sandy's lines on my own compass.

We had five lines to do that day, about seven hours of compassing, composed of five consecutive plots interspersed with straight line compassing. It was not a closed traverse, meaning that we would walk a relatively straight line of about a mile and three quarters in before having to retrace our steps out. All totaled, the distance was going to be just under four miles. It was going to make for a grueling day which is why we started out at 4:30 a.m.

The woods were light. There were no shadows, only a grey green stillness.

Low alders and shrubs were steeped in morning dew. Conifers dripped with sweet, cloying perfumed scents. The smells were invigorating.

We quickly became wet and chilled. We walked at a steady pace so as not to feel the dampness creeping into our bones. Within a few hours, we knew we would be swooning from the heat.

As the sun rose, the bush flickered with gently shifting patterns of light and shadow. A stiff morning breeze created a dappled waterfall of light on the mottled forest floor. Birds sang in the trees and ruffled their feathers.

We plodded on, stopping when needed to take our data and run our lines. Birch and poplar stands turned into Jack pine stands. Jack pine stands turned into muddy sulfurous spruce swamps.

We climbed over hills and we scrambled through dales.

These were the hardest lines we had done yet because we couldn't use the map or photos to check our bearings. There were no markers on the aerial photos such as a beaver dam or a creek or river to double check where we were. We had to place complete faith in our compassing abilities. If we were right, we should end at a very small oblong shaped pond surrounded by a dense stand of Jack pine and Balsam fir.

Noon came and went. The sun reached its zenith. The heat was stifling. Our faces were red and puffy. Our ankles were swollen and feet sore. We poured on the insect repellent to keep from being eaten alive.

We stopped for a late lunch on a rocky knoll.

"One more two hundred meter line across the ridge over there and we should come to that pond," Sandy muttered through mouthfuls of a hot and soggy cucumber and cheese sandwich.

"Man, it's going to be a long haul back too," I said, following it up with a chug of piss warm, orange cola.

Nothing more needed to be said. We lazed back on the rocks and enjoyed a half hour break.

"Think that pond will be deep enough to swim in?"

"Who cares so long as it isn't filled with leeches."

Northern Ontario lakes are full of leeches in the summer. They swim in schools a few inches under the water in the shallower lakes. I filled my hardhat with water from the middle of one lake and went to dumb it over my head when I noticed all these big black things swimming in it. Seriously gross.

This was the longest line we had done all summer without coming upon a stream or lake.

Feeling as shriveled and sun-baked as the lichen beneath our backs, we packed up and headed off, wanting to get the day over with and jump in for a refreshing swim to wash away the sweat and bug juice.

We finished our last plot and looked around.

We stood in a gully encompassed by a wall of weathered poplars and tall white spruce trees. We didn't see any Jack pine or any water of any sort.

"Let's climb that hill and take a look," I suggested.

We traipsed up the said hill which indeed offered us a pivotal view of the surrounding area.

My mind went blank.

We stood on the rocky incline looking out at the unbroken forest stands which extended each way, horizon to horizon. The forest was patched and quilted, but unwavering, not a single dazzle or sparkle of sunlight on water visible whichever way we faced.

"Oh, oh," Sandy muttered.

"Where on Earth are we?"

We both examined the maps and went over our bearings and paces. Both of these we had written down on the computer tally sheets. We came to the same conclusion at the same time. We were lost!

"What do we do?" Sandy asked, panic in her voice.

"What we don't do is run blindly into the bush," I said.

"You think it was magnetite?"

"If it was, we're screwed."

"Okay, let's sit down and go over our bearings and readings once more," I suggested.

We sat on the ground cross-legged and did just that.

"They're wrong," Sandy said, anger in her voice. She took her compass and compared the base map with the aerial map, paying particular attention to true north.

"Look here, Lolly," she said, pointing out the line we had completed four hours earlier. "Whoever did these bearings changed the reference to true north right in the middle of our lines."

"So we're probably way off course," I said, both relieved and terrified at the same time.

"Right."

"If we reverse the bearings we did on the way in we should walk out to the same plot that we started from," I offered up.

"In theory."

Neither of us wanted to contemplate what would happen if magnetite also entered into the equation. If it was in any of the ridges we climbed, we could end up really lost and the chances of being found were remote. We had plot lines everywhere within a twenty mile radius on this trip. It was a huge area for rescuers to find us.

Terror overwhelmed me. I was shaken to the core. My lunch rose up into my throat.

Shaken, but determined, we agreed to retrace our steps based on the bearings that we wrote down on our notes, not what was on the map.

117

Only by sheer force of will did Sandy and I stay on course. We didn't talk about it, but I am sure that she was just as panicked as I was. The words, 'we're lost', kept tumbling over and over in my mind. The urge to run was overwhelming.

No matter how much water I drank, the urge to puke just wouldn't go away.

We could easily be lost a few yards from our campsite like the boys had done last year, walking parallel to the lake or the road without knowing it. There were miles and miles of bush all around us.

Ben had told us to stay put and wait for help to arrive if we got lost, but we knew what had gone wrong so walking out seemed like the right thing to do.

After walking for an hour and a half, we stopped to rest. Neither of us spoke. If we did, our voices would tremble and we might just give in to panic. We could each handle our own fear if we kept it to ourselves.

As I sat on a bed of springy red club moss, I wondered idly if it was going to be the last thing that I would ever see. I examined the miniature webbing of fiber and admired its simplicity. A morbid thought, but I found it beautiful all the same.

"Do you think I could get off with an insanity plea if I murder the guy or girl who did this when we get out?" Sandy idly stirred the moss into a frowning face with her finger.

"Possibly."

"Think I might try, unless it was Ocean. I could forgive Ocean, but only her."

"I don't think it was Ocean. She hasn't the seniority to fuck up this bad."

Sandy grinned crookedly.

"We'll walk for another hour, that should get us to the end of our paces. If we don't get back to the big lake by the camp by then, we'll wait until morning and start a signal fire so they can find us," Sandy said objectively.

"Okay," was all I replied.

We stood up and stretched, checked our compasses and bearings, matched them up to our notes, and then trudged over some really rough terrain. My legs and shins ached from climbing over an endless series of sharp walled ridges and my hands were shredded from the brambles and thorns that coated the crevasses.

A loud crash echoed through the forest. Something dark stampeded towards us in the bush.

We shrieked and ran in different directions.

I ran as fast as I could. I spotted a large slash pile and hunkered down behind it. A large brown blur raced by me. I waited, my heart pounding in my chest, until it had passed. I was too afraid to see what it was.

"Sandy!" I called out after about ten minutes.

"Over here," I heard from farther up the hill.

"Keep talking to me. I'll follow your voice," I yelled back.

"What do you want me to talk about?" Sandy screamed.

"Anything!"

"Why don't you sing that awful song?"

"I don't have to."

"Why?" Sandy hollered to me.

"Because I can see you," I said, walking up to the base of the hill she had climbed.

Sandy slid down the hill on her bum.

"Did you see what it was?" I asked.

"No. I was running away from it this time."

"Shucks. I hoped you might have gotten a picture."

"No, sorry."

"Have you developed any of the film you took? You must have twenty rolls to take in to the developer."

"It's still in my pack in the tent. I thought I'd get it all developed just before the summer's end party."

"That's a cool idea," I agreed. "Blackmail everyone before they go home or back to school."

We pulled out our compasses and examined the little red needle that pointed to magnetic north. We both sighed as one. We depended on that little tiny piece of magnetized metal to save us. We hoped it was true to us.

The sun had dipped below the tree line. The forest was getting darker, the shadows blacker and denser. We had been in the bush now for a good fourteen hours. We estimated the time to be around six o'clock, neither of us given to wearing watches. We didn't need one. We went to bed when the sun went down and rose when the sun came up.

A glimmer of light caught our eye. It was tiny at first, but then grew as we slogged on.

"Think that's the lake?" Sandy asked.

"Better be," I prayed.

We strode toward the glint of light on water on exhausted legs. The glimmer grew into a rainbow of light. Waves crashed softly against the shore.

We dropped our packs and dove fully clothed into the lake. We laughed and splashed water into each other's faces. It was a glorious feeling.

We swam out into the middle of the lake. Around a crescent of shore, we saw our dark green forestry truck and our fluttering tent. Home. It was a welcome sight.

Sandy splashed at me one last time.

"Lolly?"

"Yes, Sandy."

"I don't smoke."

"Neither do I," I said after a moment's pause.

We started to laugh. How on earth were we going to start a signal fire? Rub two parched sticks together? I vowed to put one of our Bic lighters in our cruising packs as soon as we got back to camp.

We swam to shore and waded out of the water. Sandy began singing Bye-Bye Miss American Pie at the top of her lungs. I joined in the chorus.

We picked up our packs and headed for our camp. I turned and looked back the way we had come, the unsettling feeling that someone or something was watching us making the hair rise on the back of my neck. The feeling was so strong that I walked backwards for awhile, not wanting to take my eyes off our back trail, not even for a minute.

A mist had developed in the forest. I swear that someone stared out of the mist at us. I shivered, feeling chilled to the bone, and ran to catch up to Sandy.

Ahead of us, the forest seemed to part before my partner, her song bending the saplings out of her way. Sandy stormed on, full speed ahead.

A mocking jay laughed at us from somewhere deep in the forest. I cringed. We don't have mocking jays in our woods!

The minute we stumbled into camp we knew something was wrong. The tent fly was open and flapping the breeze. Our cooler and boxes of dry food were moved.

Sandy ran to the tent and I ran to check on our food supplies.

"Oh, no," she cried from inside the tent.

"What?" I called, changing directions in mid-stride.

"My camera and all the rolls of film are gone and so is my wallet. Everything has been rifled."

"Crap," I said, entering the tent. Our belongings were scattered about.

I checked the pockets of my pack. My wallet had been stolen as had a gold necklace and heart that my mother had given me for my birthday. Thankfully, I was wearing my engagement ring. Normally, I wrapped it in tissue and tucked it in my wallet.

"Where's the radio?" Sandy asked, looking around.

"I locked it in the truck," I replied.

"Where's the keys?"

"Funnily enough, I put them in my cruising pack this morning. Don't ask me why?" I said.

We then checked the truck. It was still locked and our canoe was where we left it, at the far end of the lake. We'd retrieve it later when we were rested.

We grabbed the radio from the truck and walked over to the tree that we had hung the radio antenna in. The chord still dangled down the trunk, hidden by the conifer's needles. We were in a valley and the only way to get reception was to hang the antenna up high.

Sandy plugged in the antenna and was about to call base camp when I stopped her, having the feeling that we should check everything over first.

We did so.

"Half of our food supplies are gone," I said, checking our cooler. All the meats and eggs we had left were gone.

"So are most of the cans of food," Sandy added, opening the cardboard box that held the dry foods. "Oh, wait, they left us one can of beans and a can of tomato soup."

"Lovely."

"We're going to have to retrieve the canoe and pack-up," I replied, defeated.

"What a day," Sandy groaned, disheartened.

I knew she was as devastated by the loss of her cameras as I was about the loss of my heart locket.

We called into base camp and reported the thefts. We told Ben we would be back in the morning to file the police report and replenish our food supply. We were way too exhausted to trudge back through the bush to retrieve the canoe and then pack up the tent and clean the campsite, let alone make the three hour drive back to Blind River.

After a good cry and a bowl of Campbell's tomato soup, we heard the radio crackle. Channel Six was the forestry channel and it was only for Ministry of Natural Resources use.

"Blind River Base, this is Blind River Crew Five calling in with a medical emergency. Over," came the frantic sound of Deb's voice.

"Blind River Base to Crew Five, what's the medical emergency?" Ocean's voice rang out loud and clear.

Sandy and I huddled closely together to listen, instantly worried for Deb and Jimmy.

"Jimmy's broken his leg. It's really bad. I dragged him out of the bush. He's drifting in and out of consciousness. Over."

"Hang on, Debbie, I'm going to get Ben. Over."

We waited anxiously for Ben to come to the radio.

"Aren't they on a fly-in?" Sandy whispered to me.

"Yeah, I think so."

"That's bad," Sandy groaned.

"Debbie, this is Ben. Ocean's filled me in. I've got a chopper getting ready to go. We'll be there within the hour. Over."

"Okay. He's really bad, Ben. He's broken the right leg below the knee. I splinted it in order to travois him back to camp. There's no more bleeding, but the leg is getting really swollen. Over."

"If you have any ice left in your cooler, pack it around the leg, but don't take off the splint or give him anything for the pain. We'll be there soon. Hang in there, both of you. Over."

"Damn beaver dams. Over," was Debbie's last reply.

"And we thought we were having a bad day," I said to Sandy.

We sat silently staring at the radio. That could have been any one of us. It was a sobering day.

Chapter Eight

A Whale of a Tale

August was cooler. The first week brought several days of welcoming rain which ended the drought and the fetid humidity. Layers of dust were washed away. Moss, grass and leaves turned a bright vibrant green. Purple, white, pink and orange wild flowers bloomed everywhere.

It mostly rained on our four days off. I had gone back to Ottawa to help my mum with the final details for my wedding. It was fast approaching. We had to take my wedding dress in for alterations as I had put on so much upper body muscle that it was too tight. It was a gorgeous Spanish lace antique affair fit for a countess. I wasn't the kind of gal who liked lace back then, but that dress made me feel like a princess.

Mum and I also picked out the table cloths and beautiful, deep blood orange blossom roses that looked stunning against the peach coloured bridesmaid dresses. One of my bridesmaids had just informed us that she was pregnant so we had to arrange to get a panel put in the front of the dress to accommodate her expanding tummy. Like every wedding, it was evolving into a much bigger affair than planned.

I was happy to head back into the bush just to get away from the stress of all the planning. My mother was a director and producer by heart so my fiancé and I let her go to it.

The first Monday of our tour was cool and drizzly. Sandy slept in the passenger seat while I drove, her steady breathing interrupted by the occasional snore. The monotonous hum of the motor and *thwack* of the windshield wipers made me sleepy too. In my head, I sang every John Denver, Joni Mitchell and Gordon Lightfoot song that I knew in order to stay awake. The truck didn't have a radio.

I turned off the main highway and onto a gravel side road, the old 4x4 jerking in protest. No more pavement for another week.

Sandy startled herself awake with a loud obnoxious snore.

"You sure talk a lot in your sleep," I said teasingly.

"I do not."

Sandy stretched and yawned.

I pulled the truck into the campsite that we had camped at in June and parked the truck close to a picnic table. We hoped Maurice was still around. We were thrilled to be back here because the campsite had an outhouse. Yes!!!

It seemed we were the only campers here yet again; although, there was a lot of evidence that numerous others had been here over the course of the summer including half-burned logs and a few empty beer cans.

We sat in the truck and looked out at the rain pelting down on the lake. The rain droplets bounced an inch in the air after they hit the water. We couldn't even see Spider Island as the cloud cover was so low and thick.

"Well?"

"I suppose," Sandy countered.

Neither of us moved.

The engine ticked. The windows began to steam up. We stayed there, staring out at the rain. Neither of us wanted to set up a tent in this. Our packs were in the back of the truck encased by big green garbage bags, but the humidity was so high that our clothes and sleeping bags would be sodden anyway.

After about half an hour the rain eased up from sheet-like torrents to a hazy fine mist. In silent agreement, we stepped out of the cab and put on our dark green rain slickers.

The first order of business was to string a large orange plastic tarp between four trees, covering the picnic table. We pitched the tent close enough to step out of it under the tarp.

By the time we were finished, our fingers were wrinkled and we were soaked with perspiration. Our hair hung in wet tendrils about our faces.

Large rain drops coalesced on the surfaces of the tent and tarp like wobbly gelatin. The orange tarp was blindingly bright against the muddy brown ground, grey sky and lake.

We fired up the Coleman stove and made ourselves two steaming cups of black coffee and then headed back to the truck where we sat quietly reading for a couple of hours. It was too late and too wet to start cruising now.

Eventually we moved from the cramped quarters of the truck to the cramped quarters of the tent. We hoped the weather would clear up a bit overnight.

The gloomy day passed into an even gloomier evening. The forest was too wet to find kindling dry enough to start a fire. When night fell, it was absolute, the blackness almost overwhelming in its intensity.

In a way, it was nice. I was exhausted from all the travelling and running around wedding planning. Sandy was hung-over, having gone on a four day bar hopping tour in Sudbury.

I fell asleep to the sound of the rain pitter-pattering on the tent and woke up to the same sound. I wasn't impressed. It was warm and dry inside my sleeping bag and I didn't want to leave it.

Sandy poked her head outside the tent.

"Yuck!"

"What time do you think it is?" I asked Sandy.

"I don't know, probably eight or nine o'clock."

"So much for an early start."

With a sigh, we tugged on our wet boots and still damp jeans. A steaming mug of coffee and heaping plate of scrambled eggs made me feel a bit better.

"How can you eat that?" Sandy asked in disgust as I poured ketchup on everything.

"It's the French Canadian way," I mumbled through a mouthful of ketchup covered scrambled eggs.

"That is so gross."

"Tastes great though."

"Wonder where Maurice is?" Sandy mused, putting one small strip of bacon on a plate and tucking it under the table in case Maurice did show up.

"Maybe he's moved on? Maybe there weren't enough campers to make new friends with?"

"I hope those people who came in after we left didn't hurt him."

"I doubt it," I consoled her. "He's probably just curled up in a den somewhere waiting for the weather to get better."

Sandy seemed genuinely worried about the skunk. I was a lot more wary.

We unpacked the canoe and carried it down to the water. Sandy brought down the two horsepower motor and fastened it onto the stern, making sure to chain it down well. I graciously let her carry down the gas tank too since her hands were already coated in oil from the motor while I carried the paddles, lifejackets and day packs.

We putted down the lake, heading towards the two new lines that Ben had given us.

Drizzle softly fell against the calm surface of the lake and beaded the tiny hair follicles on our faces. A single brave chickadee chirped on shore. A loon gracefully floated by us. The sound of the motor was dull and hollow in our ears.

We cut the motor and drifted into shore. The canoe bumped against the bank with a sickening thunk.

Reluctantly we set off on our tangent into the misty forest.

127

The forest was surreal. I wished we still had one of Sandy's cameras to capture the colours and texture of the vibrant moss covered trees and ethereal cloud-filled valley and lake. I half expected to see a medieval knight dance his war horse out of the drifting mists in the meadows.

Sandy crashed through the bush like a stampeding elephant. She didn't seem to notice the beauty that surrounded us. She was still angry about losing her camera and film.

I kept silent. I hadn't even told my mother about the loss of my gold heart.

The one thing that I had learned about Sandy this summer was that she hated the rain. She had no patience for soggy clothes or even soggier data entry sheets. It made for double the work as we had to transcribe everything we wrote down during the day onto dry data entry sheets at night.

I penciled in the data on the plot that we had just begun. The paper ripped as soon as placed the pencil tip against it.

"Oh, crap," I said and tried again.

We finished the two lines in twice the amount of time it should have taken. Everything had to be done over and over again because of the weather. Tally sheets were wrapped in plastic in the hopes that my chicken scratch scrawls would be legible when we got back to camp. I didn't fancy doing the same lines over again. We had to do it once before when the pencil marks had sweated off the pages because of the humidity.

Thunder rolled in the distance. A strong wind suddenly blew up. It rained without raining, the wind shaking water from the trees. Treetops swayed above our heads.

"That'll just cap off my day," Sandy said, her brows furrowing into a grimace.

"What will?"

"Getting my brains bashed in by a falling tree."

"Yeah, we should get out of here," I agreed, looking up. The trees were banging against each other. Broken branches rained down from the sky.

"Peter Pan shoes?"

"Peter Pan shoes," I agreed.

We hurriedly compassed back the way we had come. Thunder crashed. The sound was deafening. Almost immediately, lightning split open the heavens above us.

Sandy stopped to take a bearing.

"We have to get out of here fast or find shelter," I screamed over the rising wind.

More branches fell from the sky. A small one bounced off my hard hat. A tree crashed somewhere in the forest. Thunder shattered our eardrums and more lightning crackled overhead.

"Holy cow!"

Lightening illuminated the forest, making the grey day as bright as a hot sunny one.

Balls of lightning played in the treetops above our heads, hovering amidst the tips of the wind lashed poplars in a sizzling cloud of light. The air was charged with electricity. The hair on our heads started to rise.

Another loud crack and another sizzling ball of lightening rolled across the treetops.

We stood rooted to the spot, unable to take our eyes off of the heavenly spectacle.

The ball lightning crackled softly as if laughing. The sound was hideous.

We ran!

Fear tore through our hearts.

We vaulted over stumps and slash.

We were out of breath by the time we reached the lake.

The canoe thrashed on its line, pummeled by the wind and the waves. It tossed violently back and forth on its tether. The lake was a maelstrom of torrential rain and three foot waves.

More thunder peeled. We covered our ears against the canon fire. Lightning flashed all around us.

Tree after tree crashed to the ground in the forest.

"We can't go out on the lake, we'll get hit by lightning," Sandy cried out.

"We can't stay here either."

"What about the rock overhang we ran by a little ways back?"

"I don't think we have a choice," I screamed over the howling wind.

We raced back to the small cliff that we had passed when we first started out.

We huddled beneath the rock overhang, praying that it wouldn't give way under the weight of the wall of water that cascaded over its lip. It was as if Nature had unleashed its own rock album, the crashing noise of the waterfall merging with the booming thunder and wailing wind in a majestic composition that rivaled the rock opera *Tommy*. It was punctuated by the din of more falling trees.

Through the wall of water, we saw stream after stream of ball lightning dance across the tree tops.

Eventually, the storm passed and we emerged shivering from under the rock overhang.

We had to navigate back to the lake over numerous freshly fallen trees.

We hauled the half-submerged canoe onto the bank and dumped out the water. We prayed that the motor would still work and it did.

We kept to the shoreline in case the storm came back.

"I always loved watching the summer storms roll across the lake at my grandparents cottage, but I've never seen anything like this in all my life," I said.

"I understand why my professors told me to never get caught in the bush in one," Sandy sulked.

We pulled into camp and dragged the canoe as far up on the bank as we could. For added safety, we tethered the line to a tree.

We sat in the tent, the Coleman lantern providing both warmth and light, copying our notes onto dry data entry sheets. Within an hour, another storm rolled in. It was even more violent than the last.

When a bolt of lightning struck perilously close to the campsite, we ran to the truck, tucking ourselves safely away inside. The truck was far enough away from any trees that we didn't have to worry about getting crushed and the rubber tires would protect us if we got hit by lightening.

By dinner time, the weather had subsided and we were able to cook up some chicken and rice. A simple supper had never tasted so good.

☙ ☙ ☙

The next three days proved uneventful. The heavy rain turned into a drizzle and the drizzle turned into a cloudy day. We were happy to rise to glorious sunshine on the fourth day.

The temperature soared into the mid-nineties and the forest floor became brittle once again. The humidity reached 90% without the rain which made cruising sticky and oppressive.

We started getting up with the grey dawn again, about five a.m. by this time of the summer. We did the long plot lines and heaviest workload in the morning, keeping one or two short lines for the evening. That gave us the hottest part of the afternoon to swim and suntan.

Today as we left the camp, the birds were singing and several merganser and wood ducks swam lazily by our canoe as we paddled away from shore. They ruffled their feathers and quacked as we drifted past.

We fired up the motor and cruised down to the northern most tip of the lake. We had four plots to do that would take us about six hours. We seriously annoyed a heron stalking its prey in the shallows when we parked the canoe in a thicket of tangled marsh a hundred feet from its fishing ground.

It was noon when we emerged from the bush, hot and sweaty and hungry. We pushed the canoe off its weedy bank and chugged back down the lake to its midpoint where a flat plateau of rock jutted out over the water. There was a wide ledge at its base that we loved to sunbathe on. It was an ideal swimming spot too as it was about thirty feet deep there.

"I have to have a swim before lunch, what about you?" Sandy asked as I docked the canoe.

She leapt onto the rocks and held the canoe for me.

"I'm with you."

We stripped naked and went for a skinny dip. The water cooled our blistering skin. We had only the faintest of a bikini lines as Sandy and I sunbathed in the nude. I only wore my bikini at my parents home.

We swam like mermaids, splashing and dunking each other under the water. We laughed and climbed back onto the rock. We laid back, our bodies cool and whiled away the afternoon. After dinner, we would head back out to work for a couple more hours.

The canoe scraped against the rock face as it gently rocked back and forth in the faint breeze. The sun glared off of its metallic surface creating shimmering heat waves. I glanced at it, thought about tying it up, but didn't really care if I had to swim for it on such a glorious day.

Sandy sat up and rested her back against the warm rocks. She closed her eyes and went to sleep, letting her body absorb the warmth of the rock into her tired muscles. A smile crept over her face.

I laid down on my stomach. The sun warmed my own muscles and flesh. I too closed my eyes and drifted off to sleep.

A soft splash and whir from a cruising motor startled us awake.

"Hello," Sandy said.

"Good day, ladies," came the masculine reply.

I opened my eyes and glanced up at Sandy. She was still sitting spread-eagled, propped up against the rocks. She looked down at me and shrugged helplessly. She was quite unperturbed.

"Catch anything?" Sandy asked politely.

"Nope," came the brief reply.

"Good luck," Sandy finished.

"Much obliged," said the fisherman.

I turned my head without getting up and smiled at the two fisherman. They neither laughed nor grinned. The older white haired gentleman in the front of the boat sat facing us as they trolled by. He nodded hello to me as they passed. I nodded back. The younger lad, his son or grandson, I wasn't sure which, turned his back on us. Every couple of minutes though, he turned his head back around to gawk.

Sandy nonchalantly slipped into the water. I rolled off the ledge daintily and submerged myself in the lake as deeply as I possibly could without drowning.

We swam to the canoe and retrieved our coveralls, which we had left neatly folded on the seat, after the fishermen were out of sight.

"How're we going to get dressed? They're almost at the end of the lake and will be coming back in a minute," I said, embarrassed to my core.

"I suggest we climb out on the rocks and just do it."Sandy winked at me and grinned.

We climbed out of the water, trying not to laugh too loudly. For a moment, we posed on the rocks, two naked girls, bronzed as Apollo and sturdy as Amazon women, before stepping into our coveralls.

"Well they sure caught something, didn't they?"

"Two mermaids," I said lightheartedly.

"It'll be a whale of a tale in the pub tonight," Sandy said with a shake of her sun-reddened locks.

We giggled and sat down on the rocks, trying to maintain our composure as we awaited the fishermen's return.

"Have a good day," Sandy waved as they passed by us.

"Sure will," said the older gentleman.

"Sandy," I said whispered, watching them go.

"Yes, Lol."

"You do realize that we have Ministry of Natural Resources stamped all over our canoe and truck back at the landing."

"I do."

"Think we should warn Ben," I asked her.

"Why ruin his day," Sandy answered.

We laughed until we couldn't laugh anymore.

When we radioed to report in at six pm, the news of the two mermaids at Flack Lake had spread far and wide. Ben himself answered our call. He was barely able to contain his laughter.

Chapter Nine

Flip, Rattle & Roll

"Much as it goes against my better judgment, you two are getting the Terrajet for the next two weeks," Ben had said just before we rolled out of the yard for the last of our tours before I left to get married. "You'll be based out of the park."

I think Ben was being kind to me as he was the one who had taken my mother's frantic call: the wedding shop where my wedding gown was being altered had burned down and our wedding license had expired because my fiancé and I had taken it out before we headed off to our summer jobs.

It had been a rough four days off. My fiancé was on a large fire in Sioux Lookout and he didn't even know if he was going to make it out of the bush in time to get a new marriage license. I needed a new wedding dress. Were we going to make it? I wasn't sure.

I had also missed seeing Terry Fox run through both Blind River and Sault Ste. Marie. That just tripled the disappointment.

Did I say I was having second thoughts? Yeah, guess I was.

Heat rippled through the air and the asphalt melted the soles of our sneakers as we climbed into the truck. Ben had already tied down the Terrajet and hooked up the trailer for us.

"Oh, we're in for some good times, buddy," Sandy said, patting the dented sides of the Terrajet.

That didn't lift my spirits.

Sandy double-checked the chains and hookups while I started the truck.

"Want me to drive," she asked, concerned. "You don't look so good."

"Yeah, okay," I said, shimmying across the bench seat.

"Seriously, are you okay? You seem really down."

I hadn't told her about the wedding shop burning down or our lack of a wedding license. I decided I really didn't want to talk or think about it for the next ten days. We only had two weeks left anyway so I was going to work straight through.

"I don't want to die," I joked.

"Oh, heck, we're going to have a ball. Relax," she said, barreling out of the yard in a hail of gravel.

"That's what I'm afraid of."

Sandy drove at her usual break-neck speed. It was a miracle we didn't get any speeding tickets and that we had only been pulled over the once.

After a while, the hum of the highway put me to sleep.

"DAMN!"

I woke in mid-snore.

Sandy's hands strained on the wheel, her face was deathly pale. The truck swerved right and then left. The wheels shrieked in protest.

I looked over my shoulder. The trailer and Terrajet were swinging wildly back and forth.

Sandy swore again, her face a mask of concentration as she fought to keep the truck and trailer on the highway without careening down the one hundred foot embankment to our left.

Slowly, she eased the truck over to the side of the road. The highway coiled in a long snaking ribbon in front of us. The trailer skidded sideways onto the thin gravel shoulder. Thankfully, no cars approached.

I jumped out of the cab when the truck came to a jerky halt and ran for it. I thought the truck was going to slide down the embankment.

It was déjà vu all over again. The hitch sat underneath the bumper, the safety chains being the only thing that had kept the trailer from flipping. I thanked the Gods of Steel for saving our butt a second time.

I examined the trailer hitch closely. I could see that the hitch was slightly bent at the ball. No wonder it flew off. Combine that with Sandy's speed and we were seriously lucky that we weren't killed.

I walked over to check on Sandy.

Sandy's hands were wrapped around the steering wheel. They shook perceptibly. She rested her head on her hands for a minute.

"You aren't going to puke are you?"

"I might," she said.

"Let's hope that trouble isn't going to follow us in threes. That's the second time we've almost rolled a trailer." I was angry. Sandy needed to slow down.

"Yeah, I know. You can drive now," Sandy said shakily.

"I drive from now until we're done. You're a frigging maniac behind the wheel, but it wasn't completely your fault. The hitch is bent. We have to drive a lot, and I mean a lot, slower."

"Fine."

"Fine."

Together we refastened the hitch and were on our way, this time with me at the wheel.

We arrived at the park without further incident and checked in with the main office.

The only problem with bunking at the subsidiary park was that we had to work regular hours because meals in the cookhouse were at specific times. Breakfast was at 8:00 a.m. and dinner was at five. The indoor plumbing and shower were nice touch though, as was the bed, even if it was just a cot.

We kept breakfast the next morning to a minimum, knowing that we were going to be bouncing around in the Terrajet most of the day. We settled on one pancake and two thin strips of bacon a piece. The cook voiced her disapproval until we explained why.

We didn't report the incident with the trailer. We'd let them know in the yard when we returned it. We only had fourteen days to go as a crew and decided to just get on with our jobs.

We drove away from the park quietly. I kept an eye on the trailer using the side mirror, still not trusting the flat bed's hitch. Our sturdy red canoe was tied atop the metal carrier in the truck bed. Sandy yawned beside me.

I switched the truck to four wheel drive as we started to climb a steep gravel road. The truck labored up the hill, the trailer and Terrajet bouncing along behind it.

Sandy placed her hand against the ceiling to steady herself as we bounced from side to side.

I popped the truck into a lower gear. Gravel spat out from beneath the tires, the roadbed dry and slippery with loose gravel. The truck's big V8 roared. We spun our wheels half way up the grade and the truck jackknifed.

I threw the gears in park and we got out to survey the situation. Our coveralls hung like flapping duck tails around our waists.

"I don't know what we're going to do to turn around half way up the hill with the trailer on."

"We're going to be here all day at this rate," I said. I knew I sounded whiny and just didn't care at this point.

"Want to make out?"

"Funny," I growled.

Sandy laughed and went around back of the truck and unhooked the trailer.

"What are you doing?" I asked, uneasy.

"It's just the trailer that's keeping us from getting to the top of the hill. I'll drive the truck up and turn it around and we'll winch the trailer up. Watch, it'll work, I'm sure of it."

I didn't have any other ideas so was willing to try anything.

Sandy jumped inside the truck and ground the gears until she and the Dodge disappeared over the top of the hill. It only took several minutes without the trailer holding the 4x4 back.

I could hear the sound of the roaring engine and the squeal of tires as she spun circles in the gravel on the far side of the hill. I had to admit, Sandy was predictable.

I dangled my legs and boots over the side of the trailer waiting for my partner to come back. It took her about fifteen minutes to get the 4x4ing out of her system. When she was done, she roared to a screeching halt a few feet in front of me.

"I can't get by the trailer and the truck's winch is seized," she called through the open window.

"No, kidding," I mumbled under my breath. I wasn't surprised that she had tried to use the winch when she was out of sight of me either. "Turn the truck off and let's try to turn the trailer around ourselves. If we can push it over, then we can hook it up and drive it down the hill."

We each picked up an end of the trailer hitch, but weren't strong enough to lift it, not with the Terrajet on board.

"We're going to have to unload the Terrajet and drive down to the bottom of the hill. We should be able to lift it then," Sandy offered.

I agreed and climbed into the Terrajet. Sandy loosened the ties and removed the wheel blocks. Immediately the 'Jet rolled off the back of the flat bed. I gunned the motor and drove the machine down the hill, parking it well out of the way of the truck and trailer if they rolled down the hill.

Again the two of us tried to lift and turn the trailer around to no avail. We could lift the hitch, but the trailer was stuck in a rut and no amount of grunting or heaving on our part was going to budge it.

"How about we hook up the Terrajet's winch and wrap the line around the trailer hitch. Then all we have to do is let the winch pull the trailer around for us," she suggested.

"Should work. Let's give it a try."

I had to give it to her: Sandy had some good ideas.

Sandy and I walked down the hill. Sandy hopped in the 'Jet. She started the winch's motor and pushed the lever to 'Release'. I walked the cable up the hill as it unwound.

I hooked the grapple hook to the trailer while Sandy moved the Terrajet into position behind it.

I quickly realized that we needed to have more bang for our buck and unfastened the grapple hook. I stretched the cable around a tall tree and then refastened it to the trailer, using the tree as a right angled pulley.

"Let her rip," I yelled to Sandy.

Sandy started the winch rolling. The cable tightened.

I jumped out of the way as the cable snapped taught with a loud 'whack'. The hitch dug a huge gaping trench in the road bed as it slowly pivoted in a circle. It sounded like a rock stuck in a compactor.

"That should do her," I yelled, the trailer now facing downhill.

Sandy let out some slack on the cable and I let the grapple hook fall to the ground. Sandy reeled in the line and waved at me. She then jumped in the Terrajet and took off down the road.

Eventually, like a naughty child, she was done playing around and drove back.

We took out a handsaw and cut down a couple of saplings on the side of the road before driving the truck around the trailer. It only took a few minutes to hook the trailer back up.

We parked the truck and trailer at the bottom of the hill and off to the side. This snafu meant we were going to have to travel a couple of extra miles in the Terrajet. I was the only one of the two of us that found that unsettling.

🐝 🐝 🐝

I leaned into the seat back, my feet pressed against the dash to keep from getting bounced right out of the Terrajet. My stomach ached from all the bumping and jostling. The noise was painful and my head throbbed with the incessant roar of the engine.

It took us about an hour to reach the small dirt track from which bearings would be taken and we could go and do our lines.

Sandy switched off the ignition and climbed out of the 'Jet when we reached our starting point.

My legs felt like rubber. My first few steps were as wobbly as a new born foal's.

Our pace slowed down perceptibly after the second two hundred meter line. Our coveralls were soaked with sweat. Deer flies hovered around our heads, their buzzing nearly driving us mad. We stank so bad that I was surprised they were bothering us.

At the end of line, we heard the unmistakable sound of water cascading over rocks.

It took less than a few minutes to find the spring that tumbled twenty feet out of a rocky wall that ran for about three hundred paces east and west of us. The waterfall fell into a sandy pool nestled amidst the rocky base. An oasis of sphagnum moss and lacy ferns grew in abundance around the pool. Rainbows of light reflected in the mist that gathered in the crevasses.

Sandy ran fully clothed into the pool and stood beneath the waterfall.

"Man, that feels so good," she cooed, rubbing her face with the cold spring water.

I took off my stained coveralls and sat in the pool in my underwear and salt stained camisole. I used a wad of sphagnum moss as a facecloth and washed the sweat from under my arms and the dirt from my face before slipping into the water.

Sandy stripped down and rinsed her coveralls off, and then stretched them out on a sunny rock on the far side of the pool. I followed suit.

We played in the water for awhile, knowing that this was probably the last time that we would get a chance to do this.

"What time do you think it is?" I asked casually. We had at least a half hour walk back to the Terrajet and another three quarters of an hour back to the truck and then an hour to the park. Dinner was at five and the cook didn't hold it for anyone. Ingrid would have, but this cook had no patience for the cruisers.

"Time to boogey. It's three o'clock," Sandy said, checking her watch. She had taken to wearing one now that we were on a time line.

"Whoa!" I said, bustling out of the pool.

We quickly got dressed and raced back to the Terrajet. Deer flies dive bombed our wet hair, getting caught in the dripping tangles. It was tough to pull them out without getting bitten.

We reached the Terrajet and tied our packs in the back with bungy cords.

Sandy got into the driver's seat and fired up the engine. We had settled on a you-drive-one-day-and-I-drive-the-next routine. Today was Sandy's day to drive.

"You think Peter Pan could lift a Terrajet off the ground and fly with it?" Sandy asked.

"Let's find out," I laughed. "What the heck!"

"Really?"

"Really."

Sandy grinned and shifted the gear shift into low. We vaulted down the road like a runaway train.

I held onto the top of the windshield with one hand and the roll bar behind my seat with the other. It was the only way to keep from bouncing out. I still wonder why there weren't any seatbelts?

Our hair whipped out behind us, drying in the wind. We sent a deer bounding into the brush as well as several dozen rabbits. Grouse scattered in waves as we raced by them at forty miles an hour.

I glanced at my partner. Her eyes were gleaming and a wide grin spread across her face.

"Maybe we should slow down a bit," I cautioned.

She shook her head 'No' and smiled at me. Her smile reminded me of Jack Nicolson's in 'One Flew Over the Cuckoos Nest'.

We turned right onto a small rocky dirt trail. This was the last leg of our journey. Sandy continued with her foot to the floor.

Suddenly, the wheel leapt out of Sandy's hands. It spun madly. The right front tire hit a rock. The 'Jet slid sideways, first right and then left. The left tires went into a deep rut and the Terrajet flipped over.

I screamed as we rolled.

Sandy was tossed sideways over the windshield and thrown clear.

I hit the ground hard, the Terrajet rolling onto its back and pinning my legs to the ground.

"Laura, are you alright?" Sandy yelled, scrambling over to where I lay writhing in agony.

"That's a stupid question," I seethed.

She reached up and turned the key off. The whining of the engine stopped instantly. I was grateful for that small mercy.

"Help me roll this thing off me," I growled. My leg and pelvis were screaming with pain. I prayed that the pain and the fact that I could wiggle my toes inside my boots meant that nothing was broken. My left shoulder too was throbbing where it had been driven hard into the ground.

Sandy kneeled down and pushed against the top side of the machine, while I sat up as best as I could and used my legs to push upwards from underneath. The Terrajet rolled onto its side easily. I dragged myself out from beneath it.

"Can you stand?" Sandy asked, her voice not steeped in concern, but of controlled anger.

"What the hell are you pissed at me for?" I came close to shouting. "Yes, I can stand and I told YOU to SLOW down."

I struggled to my feet unaided.

"Come on," she said, softening. "Let's try to flip it back over."

We pushed against the top of the passenger door and the Terrajet rocked backwards onto its wheels with a loud bang.

"Let's hope it starts," I growled in misery.

I looked the machine over. There were several new dents and dings, but otherwise no big gouges. I picked up the gasoline can and packs from off the road and tied them down once more.

Sandy climbed inside, crossed herself, and then turned the key. It took a couple of tries, but the 'Jet roared into life.

I slowly eased my aching body into the passenger side and we continued up the track at a much more sedate pace. Every jar hurt my shoulder, hips and legs.

We didn't discuss or report the incident. One of the other cruisers called me on it when she saw the deep black, blue and purple bruises on my shoulder and left leg when I got out of the communal girls shower the next day. She urged me to report it or go see a doctor. I should have, but I didn't want Sandy or I to get into trouble, not with just a few days left.

I ended up developing blood clots in the veins in the left leg about ten years later.

🦢 🦢 🦢

Rutting season was almost upon us and we saw lots of moose scat on the roads and in the shallows around the lakes and ponds. The bulls were on the move.

We had always seen a lot of bear scat, but hadn't run into any bears since our run-in with grandpa. Sandy whined about that, but I was relieved.

We had just completed a short line off of one of the smaller nameless lakes close to the camp when we spotted a moose standing in the water not twelve feet from shore. He was about one hundred yards from our position. Our little old canoe was pulled up on the bank in front of us.

Sandy had bought a new camera and carried it with her everywhere.

"Look! Just look at him," she said enthusiastically, pulling her camera out of her pack.

The moose was young. He had a small rack, only with about three points on either side, but he was still impressive as he ate the luscious grasses growing along the shore.

"I want to get a picture of him," she said.

"Don't get too close," I urged her.

"Yeah, yeah," she said, ignoring me.

I hung back, but she crept closer. The moose looked up. He snorted when he saw Sandy, but didn't make any threatening moves.

"Hurry up, you're going to piss him off," I said, getting more and more annoyed.

Sandy snapped a few pictures and then came back to the canoe, her face glowing.

"Okay," she said, "let's go."

We threw our packs into the canoe. Sandy climbed into the front. I was about to shove the canoe off from shore and jump into the stern when Sandy let out a long low imitation moose call.

"What the heck are you doing?" I shouted.

The bull moose answered.

Sandy did it again.

I froze.

The moose called again and then raced towards us with staggering speed. Poplars and saplings were crushed beneath his weight.

"You idiot!"

I pushed the canoe away from the shore with all my might and leapt into the back of it.

"Paddle!" I screamed at a laughing Sandy who picked up her camera instead of her paddle.

The moose crashed into the water behind us as I dug the paddle in deep and starting paddling frantically.

Sandy cried out in fear, finally realizing the danger we were in. She dropped her camera, picked up her paddle, and dug in.

"Faster!" I yelled with a quick glance over my shoulder. The moose was swimming behind the canoe and he was gaining on us.

"Paddle for your life, you idiot!" I screeched.

We paddled fast and furious, slowly pulling away from the tiring moose. He followed behind us all the way into the middle of the lake.

We continued to paddle hard until we reached the boat launch where our truck was parked.

We stopped and looked back. The moose had finally given up and swam to shore.

"What were you thinking? Actually, were you even thinking at all?" I asked my partner angrily. I was so fed up with her putting my life at risk. My shoulder and left leg were still throbbing and mottled with bruises.

"I got some great shots," she said with a grin.

I shook my head, giving up. She wasn't going to change. I was glad that tomorrow was our last day together. I had a wedding to worry about and still didn't know if my fiancé was going to make it or not.

Chapter Ten

Until We Meet Again

On the last night before we left the main bunkhouse in the park, all the cruisers and park staff gathered together for a bonfire and end of summer party.

As the first stages of the glittering twilight settled in, the huge bonfire leapt at the reddening sun like a snapping dog. The orange flames stood out brightly against the darkening sky, lake and forest beyond. Sparks chased the stars in the heavens as they came out to play. A red harvest moon peeked over the horizon.

I sat staring into the flames and then out at the wide expanse of water. Dancing ripples played with the reflection of yellow, orange and red hues in heaven and earth. The scene was mesmerizing.

Sandy was partying up a storm behind me, her voice booming over the others' soft chatter. I looked over at her, her face was red from the heat from the fire and a couple cans of beer. I swear I saw the air vibrate when she spoke.

Someone brought out a guitar and the real partying began.

I turned my attention back to the lake. In the distance I could see something bobbing in the water, moving across the lake at a slow and steady pace. I watched for a couple of minutes, wondering what it was. It was pointy like a cigar and left a small wake in its path. We weren't in Loch Ness so it wasn't the fabled Nessie or Ogopogo from the Okanagan.

Sandy sauntered over to me and offered me a bottle of Canadian. The beer was wet and warm, but tasted good none-the-less. She nudged me out of my reverie.

"What are you looking at?"

"Look across the lake for a minute. In the middle. What do you see?" I pointed out the moving target.

"Looks like a moose maybe," she said, her eyes straining against the dimming light.

"That's what I was thinking."

"Only one way to find out. Let's go grab a canoe," she said, putting down her beer.

"No. Not without runner lights and permission."

"Come on, Laura, live dangerously, one more time."

I laughed and followed Sandy down to the shore where a square stern canoe sat bobbing in the dark water. We climbed into it and started up the motor, and then idled slowly out into the lake.

We putted into the middle of the lake towards the swimming figure, the red moon watching over us.

"Don't get too close or we'll scare him," I cautioned as we got closer. It was a swimming moose after all.

"I won't," she said, cutting the motor a few minutes later.

We glided to within fifty feet of the big bull moose.

"Wow, look at the rack on him," Sandy whispered.

"Fantastic," I said, my voice quavering. I was awestruck.

The huge bull's eyes glinted in the twilight. He eyed us dubiously, the whites of his eyes rolling towards us, acknowledging our presence and that of the canoe. He snorted. Water frothed around his nostrils and mouth. His breathing was heavy. He coughed out a lungful of water and picked up his pace.

We paddled parallel to him, matching his pace, but keeping well out of his way.

"If he turns towards us, I'll turn on the motor and we'll get out of here," Sandy said. This was the first logical thing she had ever said to me in the four and a half months we had worked together.

The partiers on shore were unaware of the surreal scene taking place out on the water: the swimming bull moose, the liberated canoe, and the two awestruck women inside of it.

It wasn't far now.

We followed him, keeping pace, but not approaching, until his hooves hit rock and he heaved his massive glistening bulk onto shore. He turned towards us and bellowed in defiance. With a slap of a wet tail on buttocks, he trotted off into the shadowy forest.

We watched his retreating form, letting the canoe drift wherever it wanted, until we couldn't hear his hoof falls anymore.

Night fell. The stars stretched from horizon to horizon, their glow reflecting off the flat lake. The moon's glow turned white. It hovered over the water, lighting up our way.

We paddled back to the camp, lost in the beauty around us. It was a great way to spend our last night together. The sound of the party grew as we approached, bringing us back to reality.

Over the singing and the steady flow of beer, Sandy and I glanced at each other now and then. We smiled. For tonight, the mishaps of the summer were forgotten and we were glad to have shared in each other's adventure.

We arrived in procession at the Ministry of Natural Resources office in Blind River about noon the next day. Sandy and I didn't talk much on the drive back. Our jobs were done. We may or may not ever see each other again.

Several of us exchanged addresses. I used my parents address in Ottawa as I didn't know where I was going to live in Sault Ste Marie. My fiancé and I needed to deal with the wedding first.

I was heading straight back to Ottawa in the morning. The wedding was on. The Ministry was flying my fiancé in by special plane on the Thursday. Friday we would pick up our new wedding license and on Saturday we were to be married.

We arrived at the compound and signed our equipment in including the Terrajet, canoe and truck.

Sandy was staying on to work for another month or two. Ben told us she and Debbie were to be partners. I wanted to be a fly on the wall during their first week together.

As it turned out I was right. About a month later, I received a congratulatory card from Debbie a p.s. added: *She's crazy. How did you do work with her all summer?*

We had one final party. It was fitting that we had it at The Algonquin Hotel. The Olde Country Boys were playing. We partied hard until one a.m. when they kicked us out.

The Algonquin was where I first met my cruising partner and the last place where we said our goodbyes. We never saw each other again, but that's alright, I kept a diary.

Lady of the Forest

Oh, Lady of the Forest,
how wonderful you are.
I see you on your charger,
like a shining star.

Down through the forest glade you ride,
like a knight in shining armor.
To protect all life and creatures,
Who in the forest dwell.

Through the homing mist you go,
glistening in the sunlight glow.
Ride on, oh, Forest Ranger,
and keep the forest right.
Preserve all nature's beauty,
both through the day and night.

Written by my Uncle Charlie and presented to me upon my graduation in Forestry. God bless to all.

www.ingramcontent.com/pod-product-compliance
Lightning Source LLC
Chambersburg PA
CBHW060034210326
41520CB00009B/1122